Science and Society in Modern Japan

Selected Historical Sources

The M.I.T. East Asian Science Series
Nathan Sivin, general editor

1. Ulrich Libbrecht, *Chinese Mathematics in the Thirteenth Century: The Shu-shu chiu-chang of Ch'in Chiu-shao*
2. Shigeru Nakayama and Nathan Sivin (eds.), *Chinese Science: Explorations of an Ancient Tradition*
3. Manfred Porkert, *The Theoretical Foundations of Chinese Medicine; Systems of Correspondence*
4. Sang-woon Jeon, *Science and Technology in Korea: Traditional Instruments and Techniques*
5. Shigeru Nakayama, David L. Swain, and Yagi Eri (eds.), *Science and Society in Modern Japan: Selected Historical Sources*

SCIENCE AND SOCIETY IN MODERN JAPAN

Selected Historical Sources

Edited by
Nakayama Shigeru
David L. Swain
Yagi Eri

The MIT Press
Cambridge, Massachusetts

Published in North America by

The M.I.T. Press
Cambridge, Massachusetts, U.S.A.

Library of Congress catalog card number: 74-10291
ISBN 0-262-14022-5

Printed in Japan

Contents

General Editor's Foreword

The M.I.T. East Asian Science Series

One of the most interesting developments in historical scholarship over the past two decades has been a growing realization of the strength and importance of science and technology in ancient Asian culture. Joseph Needham's monumental exploratory survey, *Science and civilisation in China*, has brought the Chinese tradition to the attention of educated people throughout the Occident. The level of our understanding is steadily deepening as new investigations are carried out in East Asia, Europe, and the United States.

The publication of general books and monographs in this field, because of its interdisciplinary character, presents special difficulties with which not every publisher is fully prepared to deal. The aim of the M.I.T. East Asian Science Series is to identify and make available books which are based on original research in the Oriental sources and which combine the high methodological standards of Asian studies with those of technical history, and to bring special editorial skills to bear on problems that arise when ideas from diverse worlds of discourse are interwoven. Most books in the series will deal with science and technology before modern times in China and related East Asian cultures, but manuscripts concerned with contemporary scientific developments or with the survival and adaptation of traditional techniques in China, Japan, and their neighbors today will also be welcomed.

Science and Society in Modern Japan

This collection of essays by Japanese scientists and historians of science is the first book in the series to discuss modern East Asian science and technology. It documents the experiences, dilemmas, successes, and failures of a cosmopolitan but until recently isolated scientific community.

Emerging from self-imposed national seclusion in the middle of the 19th century, Japan has since experienced modernization by imperial fiat, a severe depression, a period of nationalistic totalitarianism in which no scientist was above the scrutiny of the "thought police," and a disastrous war. Against this background, Japanese scientists—no more inclined by nature to political reflection than their counterparts elsewhere—were forced to confront their relations as professionals to a society increasingly propelled toward tragedy but demanding absolute loyalty. History, political theory, and sociology thus became essential tools for understanding their situation and creating organizations which would give them a common voice.

The introductory section describes how Japanese historians of science view their own scientific heritage and explain the present Japanese tendency to focus on social priorities and needs as they influence scientific change rather than on the "internalist" history of scientific ideas which is still prepondent in Western Europe and the United States.

Part I presents several milestones in the formation of a characteristically Japanese ideology of science, from Ogura Kinnosuke's pioneering social analysis of Renaissance arithmethic (1929) to recent writing on the future of a technological society. Part II, a typical spectrum of postwar writing by contemporary historians of science, provides food for reflection on how historical concerns have shifted and been redefined with changing social circumstances and with the gradual shaping of the study of scientific change as a professional discipline. Part III brings the two earlier strands together in a group of discussions by scientists and historians of scientists' movements in Japan and their professional, institutional, and wider social backgrounds.

The need for a representative and competent book on modern Japanese science and its relations to social change is of such long standing that the more general significance of this one may easily be overlooked. That would be very unfortunate. People who are curious about the development of science usually begin by learning about the great discoverers, and come naturally to think of scientific activity as driven by curiosity untainted by mundane social, political, or economic considerations. Even when attention to these factors becomes unavoidable—since, for instance, most of the great discoverers of very recent times have been salaried employees—they are still considered somehow extrinsic to the real life of science. That is largely because the social, political, and economic involvements of Western scientists over most of the last century have been so conventional that they can be easily overlooked. The worldwide political and social turmoil of the last decade is leading to a gradual increase in the number of people concerned about the social matrix of science and engineering, who in fact see this exploration as a matter of some urgency. There seems to me a genuine danger that the historical study of science in the United States and Western Europe will soon congeal two mutually exclusive viewpoints—one which prefers to think about scientific change as outside society and the other convinced that scientific ideas, as the mere products of social processes, need not be taken seriously on their own merit.

Both of these points of view seem better suited to the production of mythology than of historical understanding. Both depend for their growth and maintenance on a large number of professional historians, philosophers, and other students of science who are so comfortable with their prejudices that they need not consider uncomfortable alternatives.

We badly need to begin thinking out a unitary comprehension of science which will be capable for the first time of integrating the internalist and externalist points of view, so that we can begin to understand the extremely complex relations between the growth of ideas, the influence that society exerts upon them, and their reciprocal influence. Perhaps our most nourishing food for thought will come from the study of periods when the relations of science and society were in crisis. That is what this book is about: the attempts of scientists in Japan over the last half-century to create means for dealing with uninterrupted crisis. What began as a crisis of nationalist militarism metamorphosed into a crisis of war, was transformed into a crisis of military defeat and the attendant collapse of a traditional system of values, and finally issued in a crisis of prosperity accompanied by awesome destruction of the physical and spiritual environments and a general level of poisoning in air and food that most Japanese are no longer able to cope with emotionally or intellectually.

There is an undercurrent in this book of discomfort, of recourse to the past out of concern with the present, of casting about for badly needed answers that may not exist, which has stayed with me as a reminder of how serious a business the history of science should be. This volume is no more a model of the integration of internalist and externalist approaches than any other book now available. On the other hand, I believe that the dilemmas and responses it chronicles will lead some perceptive readers closer to understanding that the humanistic study of science, when it is not merely academic (and regardless of subject it never need be), actually can be used to evolve a unitary vision of science as simultaneously an advancing approximation to truth about physical reality and a cultural artifact decisively shaped by social values.

Nathan Sivin
Technology Studies Program, MIT

Foreword

It is my great honor as president of the History of Science Society of Japan, which is host to the XIVth International Congress of the History of Science in Tokyo and Kyoto in August 1974, to present in the English language this book on *Science and Society in Modern Japan* as part of the preparations for the Congress. The papers collected herein are considered typical of the varied interests in and approaches to the relations of science and society among those most active in this field during the last few decades.

Unfortunately, only a few studies by Japanese researchers in the history of science have been accessible to non-Japanese readers because of language difficulties. Thus, most of our studies written in Japanese could not attract international attention, though some of them possess high quality and creativity.

The publication of this book is a major step toward bridging the scholarly communications gap between Japan and other countries. For the above reasons I should like to commend our editors' efforts without which this book could not have been published. Thanks are also due the Ministry of Education of Japan which provided both a publication grant and an editing grant for two years (1972–1973). The cooperation of the director of the University of Tokyo Press, Minowa Shigeo, also was invaluable in the production of this book.

Yuasa Mitsutomo

December 1973

Foreword

Introduction

Simply because of the language barrier, if for no other reason, many sectors of Japanese scholarship have thus far been inaccessible to international audiences. While technical data on Japanese developments in science and technology have been transmitted with less difficulty into international languages, few attempts have been made to translate into other languages the debates and dynamics that move the Japanese scientistis and technologists themselves. Some basic concepts like *gijutsu* and *kōgai* have no precise Western counterparts, and simply to translate them respectively as 'technology' and 'pollution' (which, for convenience, we must) obscures the fact that they form part of an independent yet still isolated tradition of unique problems and their answers.

The person who desires to be truly international in his perspective but does not read Japanese may be annoyed if told that in a linguistically-isolated island country there is a prolonged and serious intellectual struggle which he cannot directly explore very easily. Most of his Japanese counterparts, on the other hand, have been so busy trying to catch up with what is going on outside their own country that they have had no time for or interest in bringing their sense of social and intellectual struggle into the international arena.

When plans were initiated for the XIVth International Congress of the History of Science to be held in Japan in August, 1974, and especially because of a symposium on the "professionalization of science" projected within it, we were faced with the necessity of bridging the gap between Japan and the rest of the world in the way of sensing problems and in the vocabularies used to discuss them. Not overly optimistic about performing such a difficult task, still we felt something must be done to make this gap as narrow as possible, or at least to help make the symposium meaningful. One preparatory step toward that goal, we felt, was to publish in English some of the major contributions of Japanese historians of science.

This book, then, is a collection of papers considered representative of the main areas of concern among Japanese science historians. Part I consists of discussions of emergent ideologies of science as defined by some of its primary participants. Part II advances the discussion into several areas of exploratory research. Part III focusses on the concern of Japanese science historians with their social environment and some of the scientists' movements organized to express that concern.

One of the important reasons why Japanese scientists become historians of science seems to be their criticism of the internal structure of the scientific disciplines, or of scientific research as such, though not uncommonly their critique extends to the social relations of science, in what may be called the externalist approach to the history of science. Because of the importance of this approach in Japan, this book begins with Nakayama Shigeru's "General perspective," which attempts, for Western audiences accustomed to emphasis on the unfolding of scientific ideas as they relate to each other, to explain the Japanese stress on social priorities and forces as they influence the course of science.

Part I first presents one of the earliest examples of this approach, a portion of Ogura Kinnosuke's memorable article (I-1), published before the celebrated paper of Boris Hessen on the socioeconomic roots of Newton's *Principia* (and its similar impact on the West), which inspired Japanese historians of science to take up the externalist approach. Taketani Mituo's paper (I-2) gives the background of his view of science, a view which has strongly influenced the postwar generation of science historians. A paper by Hoshino Yoshirō (I-3) and a book review by Shizume Yasuo (I-4) provide outlines of a controversy over the concept of *gijutsu* (technology) characterized by uniquely Japanese interpretations of Marxism.

Part II opens with Oya Shin'ichi's article (II-1), which gives some insight into the views of Japanese historians of science and some suggestions for future work in this field. Then, Nakayama Shigeru's article (II-2) inquires into the viability of the university as an institutional base for science. From a simple methodology of enumeration Yuasa Mitsutomo in his paper (II-3) moves into some very complex questions as to the course of science in the modern world. This part concludes with Yagi Eri's preliminary studies (II-4) on quantification in the history of science.

We often note that reports by Western observers of the Japanese scientific scene tend to be based on interviews with members of Japan's science establishment and also on official records; hence, their accounts have inevitably been optimistic, if not also superficial. Most Japanese historians of science have come into this field through participation in various scientists' movements which generally maintain a critical stance toward the science establishment. To grasp the historical dynamism of these movements, some understanding of the Japanese scientists' struggle with their social context is indispensable. Part III, therefore, opens with Oka Kunio's account (III-1) of his personal experiences of the difficult prewar days of suppression of scientists' movements. Good examples of the externalist approach to institutional innovation and the formation of a research tradition are found in the joint article by Itakura Kiyonobu and Yagi Eri (III-2) and the one by Hirosige Tetu (III-3). Kaneseki Yoshinori's article (III-4) explores trends in the democratization of the field of physics, highlighting both mobility and mutual cooperation among young physicists. Nakamura Teiri (III-5) unravels a complex controversy in the field of biology to show that strong progressive social support of scientific theories does not necessarily protect them from rejection based on internal analysis and testing. And the article (III-6) by Nakayama Shigeru on the field of geology demonstrates both the attractiveness of the democratized methods and focus of a scientists' movement and their potential vulnerability under the dominant pressures of heavily equipped "big science." Finally, Ui Jun's paper (III-7) is not a mere historical account but an up-to-date report by the leader of a scientists' movement to rescue the people's interest in modern science from the more destructive aspects of state and corporation management.

While a deeper grasp of the struggle of scientists' movements and of their methodological critiques of the relations of science to society is the main focus of this volume, it may also be important (at least for any reader not particularly familiar with Japan's modern history) not to overlook some of the more obvious and direct relations between science and society in Japan. To elucidate the observations to follow, a simple historical chart is provided.

Not least important has been the impact of major international events and processes on the adoption, advancement and criticism of science. Western pressures in the mid-19th century that forced the opening of the country also resulted in the Meiji government's all-out importation of modern science and technology. Japan's own international conflicts with China and Russia, and with Western powers in the Pacific area, as well as the impact of major world conflicts like World Wars I and II, the Korean and Vietnam Wars, not to mention the Great Depression, all had crucial effects on the mobilization of resources and the direction of their use in science and technology. While earlier world pressures prompted Japan's adoption of modern science, later ones (as Nakayama's "General perspective" shows) induced some scientists to enter into the history of science as a ground for reflection on their profession. Hence, in the earlier decades of Japan's modern century, there was little work done in the history of science (except for chronology; cf. II-1), though in later decades there appeared many works critical first of society and then of science itself.

To press the obvious a bit further, it is mere commonplace to observe that Japan's modern century has had two distinct phases: the first was the initial eight decades or so of Japan's emergence as a modern state and the collapse of it in World War II (1860's to early 1940's). The second is the current three-decade period of postwar economic recovery and expansion (1945 to the present). Early in the first phase Japan enjoyed fairly full access to the sources and equipment of modern science and technology. Patent rights were purchased, machinery imported, students sent abroad in droves, and squadrons of teachers and technical experts brought in. Meantime, schools, universities and research institutes were built, learned societies formed, dictionaries and texts prepared in Japanese, and trained Japanese began to replace foreign tutors. Just as Japan was gaining some confidence to sustain her own scientific enterprise, World War I intervened to cut her off from full access to sources and equipment. Thrown somewhat on her own resources, Japan embarked on an altered, more independent course—only to be stymied again by global and domestic depressions. Ultranationalism may have been a natural response to the repeated setbacks, but was carried to excessive extremes in attacking rationalism and suppressing progressive scientists' movements. Not a few of the papers in this volume give evidence of the distortions in intellectual activity caused by the social and political strains of the 1920's, 1930's and early 1940's (and it is precisely the social history of science in Japan during this period, and the intense intellectual responses to it, that have been virtually untouched by English-language materials).

Against the truncated perspectives of Japanese nationalism, Marxism had a universalistic appeal—much more so than other systems like existentialism, pragmatism, etc., which appeared to be still too characteristically Western for application to Japan's problems. Moreover, the social history of science in the West had also been ignited by Marxist thought. Before World War II, therefore, the history of science as studied in Japan was in many respects a branch of applied Marxism. The story of Yuiken (III-1) is only the more obvious of many instances. Only those in parts of the world, particularly the Anglo-Saxon parts, where Marxist insights were long forgotten until recent discovery, will find this strange.

The social experience of wartime hardships was not, however, all bad. Total mobilization on an unprecedented national scale provided an experience of real mass production that fostered the diffusion of professional scientific and technical skills (including education, engineering, management, etc.) which were an invaluable legacy to the postwar tasks of recovery and expansion. Papers in this collection demonstrate that scientists and historians of science were among the most sensitive to whatever prewar distortions and problems were carried over unresolved or in new forms into the postwar era. Sincere and serious questioning of the present-day scientific enterprise in Japan therefore remains high on the priority list of contemporary historians of science—and serves as a check against the usual simplistic appraisals of this era as 'miraculous'.

The question of what kind of science, and how much of it, is good for society, or how much the pace of social change is a purely scientific or technological matter is not directly posed; yet, not a few of our papers contain the hint that any questioning must not be done too dogmatically, lest the ghost of excessive prewar ideologism return to haunt us. Certainly the postwar historian of science works within a more pluralistic philosophical context, and the vast increase in the number of university-educated people gives him a much larger and more knowledgeable, if also more diverse domestic audience.

This fact is not insignificant in considering problems of science and society. For much of the past century "science and society" have been conceived largely as "science and government," or in extreme instances, "scientists vs. officialdom." This reflects, on the one hand, the fact that since the Meiji Restoration (1868) Japanese science has been almost exclusively the province of government sponsorship and direction, and on the other hand, that the history of science in Japan, as a scholarly activity, developed spontaneously outside the realm of institutional support. Hence, many early proponents of science history were self-supported writers or avocational activists trained as professional scientists but maintaining a dissenting tradition in communities of like-minded critics.

The history of science today is gaining some institutional footing, but retains much of the early dissenting spirit. In postwar Japan, science is less a recipient of large government expenditures than in many advanced industrial societies, and Japan thus far has wisely avoided indulgence in some of the more wasteful sorts of scientific extravaganza, especially that promoted for purely political purposes. It would be a mistake for an outsider to assume, however, that Japan's postwar economic prosperity has been backed by a clear and consistent science policy; quite the opposite, scientific research has managed relatively well despite a *laissez-faire* setting. Of course, increasingly large Japanese corporations, and even multinational corporations, are becoming the patrons of science and technology; and to the extent that their uses of science and technology are inimical to the public interest—as seems increasingly the case—the tradition of critical dissent among Japanese historians of science will surely remain vigorous. That the foundations of this tradition have been laid with insight and integrity will be appreciated, we hope, even from the limited collection of papers presented in this book.

The editorial apparatus adopted for this book is quite simple. The reader will note, for example, that due to the large number of book and article titles, only the first word and all proper names of persons and organizations in the titles are capitalized. Titles of publications in the Japanese language appear in the text and references in italics, with approximate English equivalents in parentheses. In Japanese names of persons, the surname always precedes the personal name. (Historically there have been several systems of romanization of Japanese words, and thus a few names of persons, for example authors Taketani Mituo and Hirosige Tetu, have been established publicly in romanized forms other than the Hepburn system which generally prevails in this book.)

All three editors have checked each paper for accuracy as regards meaning, though final responsibility for the English text was assumed by David Swain. Each editor also translated one or more of the papers.

We would like to express our sincere appreciation to Dr. James Bartholomew, assistant professor of Ohio State University, for his help in preparing translations of papers II-2, III-2 and III-4, and for providing the Annotated Bibliography of English Language Works on the Social History of Modern Japanese Science; and to Fujita Chie for translating the initial draft of paper III-1, for compiling some of the biographical data on our contributors, and for arranging the Index. Again, final responsibility for the published form of all papers rests with the editors.

To the editors and publishers of the books and journals from which the papers were selected we express our gratitude for permission to use these papers in this collection, and this gratitude extends also to the authors themselves (or to the heirs of deceased authors). Many of the authors were kind enough to go over initial drafts of translations and make corrections and suggestions for improvement, and for this too we are deeply grateful.

Finally, we are most thankful for the professional skill and sustained cooperation of the University of Tokyo Press director Minowa Shigeo and of its chief editor of English Editions, Amadio Arboleda, in the production of this book.

David L. Swain
July 1973

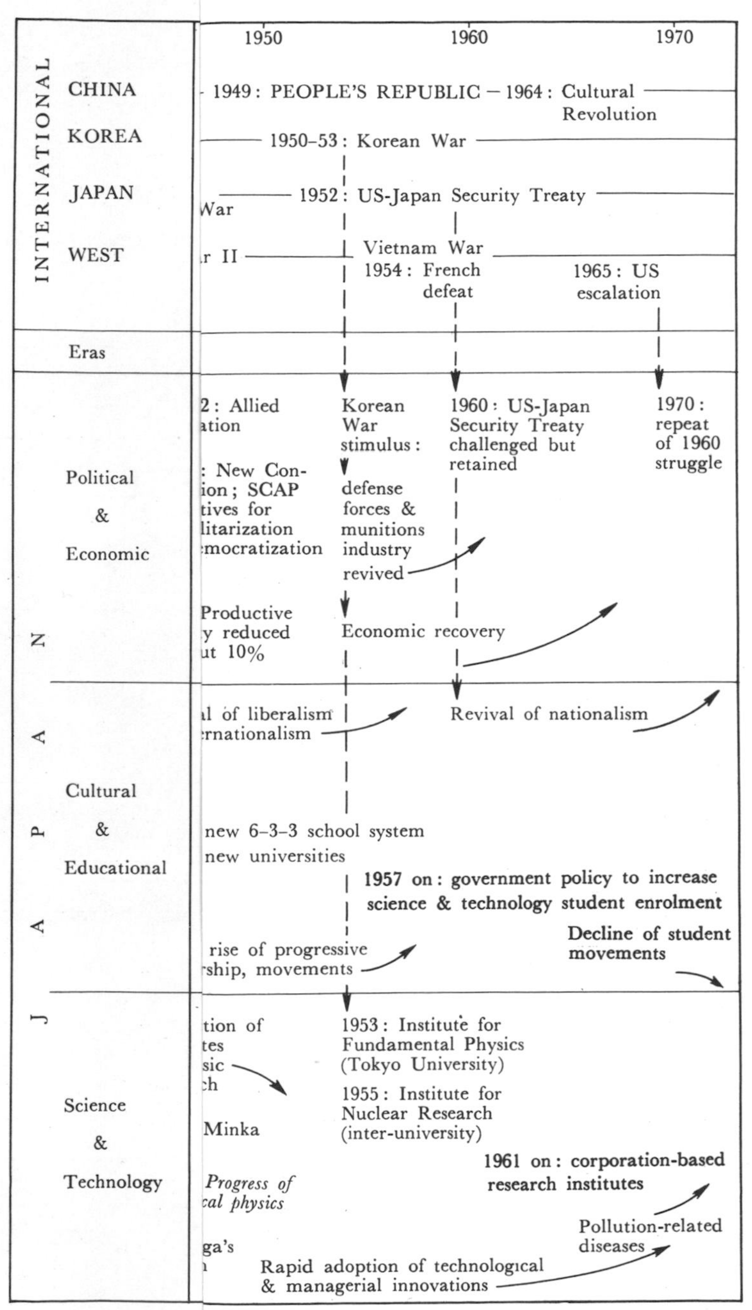
1950
1960
1970
INTERNATIONAL
CHINA
- 1949: PEOPLE'S REPUBLIC — 1964: Cultural Revolution
KOREA
1950–53: Korean War
JAPAN
War
1952: US-Japan Security Treaty
WEST
r II
Vietnam War
1954: French defeat
1965: US escalation
Eras
JAPAN
Political & Economic
2: Allied
ation
: New Con-
ion; SCAP
tives for
litarization
mocratization
Korean War stimulus:
defense forces & munitions industry revived
1960: US-Japan Security Treaty challenged but retained
1970: repeat of 1960 struggle
Productive
y reduced
ut 10%
Economic recovery
Cultural & Educational
l of liberalism
rnationalism
Revival of nationalism
new 6–3–3 school system
new universities
1957 on: government policy to increase science & technology student enrolment
rise of progressive
rship, movements
Decline of student movements
Science & Technology
tion of
tes
sic
ch
1953: Institute for Fundamental Physics (Tokyo University)
1955: Institute for Nuclear Research (inter-university)
Minka
Progress of
cal physics
1961 on: corporation-based research institutes
ga's
Pollution-related diseases
Rapid adoption of technological & managerial innovations
D. Swain July 1973

Note in Introduction

As production of this book commenced at the University of Tokyo Press, conversations with the M.I.T. Press arose as to the possibility of including this volume in the latter's M.I.T. East Asian Science Series. The Series' general editor, Nathan Sivin, professor of the history of science and of Chinese culture at the Massachusetts Institute of Technology, kindly went over the manuscript and made a number of helpful editorial suggestions. As many adjustments as possible, within the limits of our production schedule and budget, were made to conform to the style and format of this series, in which we are most pleased to have this book listed.

The Editors
January 1974

Biographical Notes on Contributors

Bartholomew, James R., born June 30, 1941 in Hot Springs, South Dakota, U.S.A., received his bachelor's, master's and doctoral degrees from Stanford University in 1963, 1964 and 1972 respectively. His doctoral dissertation was on "The acculturation of science in Japan: Kitasato Shibasaburo and the Japanese bacteriological community, 1885–1920." Presently an assistant professor of history at the Ohio State University (since 1971), he was a research fellow at the University of Tokyo from 1967 to 1969 and then a special student at Harvard University (1969–1971). His research interests center on modern Japanese history and the social history of science (see Bibliography for his publications).

Fujita Chic was born October 9, 1931 in Tokyo, and graduated from the science faculty of Ochanomizu Women's University in 1955. She worked for several years as a chemist at the Institute of Applied Microbiology, and is now a free-lance writer of children's scientific materials for books and journals.

Hirosige Tetu, born August 26, 1928 in the city of Kobe, received his undergraduate degree in physics in 1952 from Kyoto University's faculty of science, and later his Ph. D. degree from Nagoya University for his work in the history of physics. An associate professor in the science and engineering faculty of Nihon University, he is interested in both the history of modern physics and the development of scientific communities in Japan, and has published a number of works in both areas. His publications have played an important role in the popularization of the history of science, and have helped to establish a place for the history of science in the world of Japanese learning.

Hoshino Yoshirō, an active advocate of Taketani's concept of technology, was born January 13, 1922, and graduated from Tokyo Institute of Technology in 1944. A free-lance researcher and writer, he has been active for many years in exposing through his published works the deeper aspects of modern technology. Hoshino is also noted for his effective work as an organizer of movements for environmental protection.

Itakura Kiyonobu, born in 1930, took his undergraduate degree in the history and philosophy of science at the University of Tokyo in 1953, and in 1958 received a Ph. D. degree for his work in the history of physics from the same university. Author of several books, Itakura is a staff member of the National Research Institute for Education. Particularly interested in the application of the history of science to education, he has devised methods for teaching physics to school children which have been successfully introduced at many schools in Japan. Called the *kasetsu jikken* (experimental hypothesis) method, it requires pupils to predict the results of an experiment before carrying it out. With Kimura Tōsaku and Yagi Eri he has co-authored a biography of Nagaoka Hantarō (in Japanese) published by Asahi Shinbunsha.

Kaneseki Yoshinori was born February 4, 1915 in Niigata prefecture. After taking his undergraduate degree in physics in 1940 at Tohoku University's science faculty, he became a science reporter for the *Mainichi shinbun* after World War II, reporting changes in various scientific organizations from a critical point of view. Presently a free-lance researcher and writer, Kaneseki also has translated several general works on physics.

Nakamura Teiri was born in Tokyo on January 7, 1932. A biology graduate (1958) of Tokyo Metropolitan University's faculty of science, he is currently an associate professor in the general education faculty of Rissho University. A specialist in the history of biology, he is best known for his publication on the Japanese dispute over the Lysenko theory.

Nakayama Shigeru was born in Hyogo prefecture, June 22, 1928. He did his undergraduate work in astronomy at the University of Tokyo's faculty of science, and in 1959 received his Ph. D. from Harvard University. A lecturer in the faculty of general education of Tokyo University, he is the author of books on the history of astrology and astronomy as well as many articles on the history of science. His *A history of Japanese astronomy* was published by Harvard University Press in 1969. Nakayama also translated Thomas Kuhn's *The structure of scientific revolutions* into Japanese, and has made effective use of Kuhn's concept of "paradigm" in his research. He is also a co-editor with Nathan Sivin of *Chinese science*, M.I.T. East Asian Science Series, 2 (M.I.T. Press, 1973).

Ogura Kinnosuke was born March 10, 1885 in Yamagata prefecture, and died October 21, 1962. Graduated from Tokyo College of Physics in 1905, he led an active life as director of the Shiomi Institute for Physical and Chemical Research, president of Minka (Association of Democratic Scientists), and president of the History of Science Society of Japan. The author of more than ten books on mathematical education and the history of mathematics in Japan and in Europe, Ogura was a leader in the establishment of the history of science in Japan and a highly respected member of the academic community in Japan. His pioneer work on the social impact of the development of mathematics, published in 1928–1929—earlier than Boris Hessen's *Science at the crossroads*—brought the understanding of mathematics into close relation with the grass-roots level of everyday life.

Oka Kunio was born January 15, 1890 in Yamagata prefecture, and died May 22, 1971 in Tokyo. Graduated from Tokyo College of Physics in 1914, he taught at Tokyo's No. 1 Higher School until his retirement in 1931, after which he was an active essayist. Belonging to the Marxist group of Japanese scientists, Oka worked outside the mainstream of academic life in Japan. He published numerous works explaining the Marxist approach to the history of science and technology.

Ōya Shin'ichi was born November 3, 1907 and graduated from Tokyo College of Physics in 1940. A professor of Fuji Junior College, he specializes in mathematical education and in the history of Japanese science. His research centers in Japan's premodern science, and his published works have thrown light on achievements from the 14th to the 19th centuries, especially on the development of *wasan*, the indigenous mathematics developed during Japan's period of isolation (17th to mid-19th centuries).

Shizume Yasuo was born November 27, 1925 in Tokyo. A physicist specializing in biophysics and the history of science, he graduated in 1947 from the University of Tokyo's faculty of science. Shizume is well known in Japan for his translations of J. D. Bernal's *Science in history* and of N. Wiener's work on cybernetics. Formerly the editor of Minka's journal *Kokumin no kagaku* (Science for the people), he is currently working on a political philosophy which aims at peaceful coexistence between non-Stalinist communist states and reformed capitalist states.

Swain, David L., was born December 13, 1927 in North Carolina, U.S.A., and received his undergraduate and graduate training at Duke University. A two-decade resident in Japan, he has published translations of urban sociology (*The Japanese city: a sociological analysis* and *Social change and the city in Japan: from earliest times through the industrial revolution*, both by Yazaki Takeo), and collaborated with Sugimoto Masayoshi on a forthcoming joint work, *Science and culture in Japan* (premodern period).

Taketani Mituo was born October 1, 1911 in Fukuoka prefecture and received his undergraduate degree in physics in 1934 from Kyoto University's faculty of science. A scholar with wide interests in philosophy, theoretical physics, technology and the history of science, he taught at Rikkyo University before becoming a free-lance researcher and writer. Deeply involved in research and criticism of different aspects of theoretical physics, he has been one of the most influential members of the Japanese group of nuclear physicists. His innovative concept of technology as the conscious application of objective laws for productive purposes has attracted a strong following in Japan. In addition to more scholarly works on atomic energy, materialism, and technology, his book *Shi no hai* (Ashes of death) on hydrogen bomb tests had a strong impact on the Japanese public.

Ui Jun was born in Tokyo in 1932, and received his undergraduate degree in applied chemistry from the University of Tokyo's engineering faculty in 1956. An environmental protection engineer, he has published several books on public safety and environmental protection theory and practice. Active in publicizing the cause of victims of mercury-pollution (Minamata disease), he is widely known as an effective opinion leader for the anti-pollution movement in Japan.

Yagi Eri, born October 9, 1931 in Tokyo, graduated from the science faculty of Ochanomizu Women's University in 1954. The following year she entered the graduate school of the University of Tokyo and, after study abroad at Yale University (1960–1963), received a Ph. D. from the University of Tokyo in 1965 for her work on several topics in the history of physics in Japan. Especially interested in the methodology of thc history of science, she has published various articles on the statistical approach to the history of science and on Nagaoka Hantarō's Saturnian atomic model and its development. With Itakura Kiyonobu and Kimura Tōsaku, she coauthored a biography of Nagaoka Hantarō (in Japanese), published by Asahi Shimbunsha (1973). She is an associate professor in the faculty of engineering of Tōyō University.

Yuasa Mitsutomo, the current president of the History of Science Society of Japan, was born December 20, 1901 in Fukui prefecture. Formerly professor at Kobe University, he now teaches at Senshu University. He is well known for a variety of publications which have popularized the history of science in Japan. His research methods are based mainly on statistical determination of scientific activities through investigation of chronological data.

General perspective

The History of Science
A subject for the frustrated*

Nakayama Shigeru

Two generations of science historians

Among those active today in the history of science in Japan, we are immediately struck by two peaks that stand out in the age-groupings, namely, a generation born in the early 1900's who started their professional careers in the late 1920's and early 1930's (hereafter called the prewar group), and another group born in the late 1920's who entered the field after World War II (referred to below as the postwar group).

Unlike a mature academic field where some mechanism for recruitment through the higher education system has been established, a brand new discipline like the history of science has no way to assure the continuous production of its professionally committed personnel. Younger scholars have had to commit themselves to the field without job guarantees, and thus face the possibility of rejection from the established academic world. Hence, the rise and fall of the production rate of science historians has necessarily and directly reflected various external (non-institutional) causes as well as an overall *Zeitgeist*. It may not be too far-fetched to explain the emergence of these two distinct generations in connection with the two major wars of this century.

Not involved in World War I to any serious extent, Japan reaped a huge economic harvest in the absence of Western competitors. Just after that war the Japanese government in 1919 issued the "University Act" with the stated purpose of expanding higher education to match the now enlarged national prestige and economic capacity of Japan.

The pre-World War II generation of science historians en-

* Appeared first in *Japanese studies in the history of science*, No. 11 (1972), pp. 1–10, under the title "The externalist orientation of Japanese historians of science."

joyed the benefits of this Act; numerous students flooded into the expanded system of higher education. When they graduated from the universities in the late 1920's, however, the Great Depression came and a surplus of college graduates suffered from widespread unemployment. This was also the time of a rising Marxist ideological wave. As a matter of course, this generation turned out to be very socially minded and some of them were, no doubt, influenced by the Marxist approach to the history of science, as exemplified by Hessen.

This sketch of the typical Japanese historian of science belonging to the prewar group—albeit an oversimplified one—is superbly confirmed by Tosaka Jun, the leading Marxist philosopher in prewar Japan, in his article "Saikin Nihon no kagakuron" (Recent Japanese arguments on science), in *Yuibutsuron kenkyū* (Studies on materialism, No. 56, 1937) as follows:

> "For the last ten years favorable industrial and technological circumstances have led to an increase in the number of young intellectuals who choose careers in science or engineering. This is evident in the number of students in higher schools who major in the sciences and humanities. However, even an expanded industrial sector has not in fact been able to absorb sufficiently the rising generation of young engineers and natural scientists. At least this was the situation before the recent passage of the new defense budget, narrowly defined. Hence, young engineers and natural scientists have faced a kind of unemployment problem. Actual unemployment occurs only in extreme cases; in most cases graduates form a quasi-professional class in the natural science field, that is, a reserve supply, or indeed, a congested surplus of those seeking legitimate professional status. This quasi-professional group, unlike established university scientists before them, have been exposed not only to training in post-World War I social thought but also to the various social contradictions directly affecting their personal livelihood and future job prospects. Therefore, they are naturally led to play the role of bringing natural science, heretofore possessed exclusively by bourgeois academics, into the context

of social thought. . . . Thus it is that, despite the limitations described above, the quasi-professionals in natural science are destined to perform an extremely useful social function. Their participation in the 'philosophizing' of, or scientific examination of research in the natural sciences is, whether conscious or not, an outcome of this function."

The next occasion of encouragement for scientific careers came just before and during the World War II effort to meet the wartime shortage of scientists and engineers. During the war science majors were regarded as reserve scientific manpower and enjoyed the privileges of exemption from, or postponement of compulsory military service, while students majoring in the humanities were called into service and nearly eradicated from the campuses.

When the war ended the surplus of scientific manpower found it difficult to obtain regular scientific jobs, as the whole industrial sector was closed down, university laboratories had been destroyed by sustained bombing, and equipment for carrying out scientific research was not available. Again, in the late 1940's, a generation of frustrated young scientists turned out to be very socially conscious. Such were the shared circumstances of science historians in the postwar generation, to which the present author belongs, though individual motivations to give up a normal career in science and turn instead to its history are complex and varied, and it is not easy to generalize about them.

History of science: a subject for the frustrated

Anyone who is content with the practicing of science, more or less taking for granted the conventional value of the professional community to which he happens to belong, can be a good scientist but not always a good critical historian of science. When such a person turns to writing a history of his profession, the work usually turns out to be a self-congratulatory narrative of the orthogenetic evolution of his discipline.

Only for those who fail to conform or otherwise accommodate themselves to the norms of the existing scientific community does the gap between one's original image of science (or something within that image to which he cannot conform) and the existing community become the vital source of a critical attitude toward the practice of the contemporary scientific profession. Some of them may seek possible alternate courses for the development of science by returning to historical origins, or by examining the later points of divergence precipitated by particular choices as to developmental courses. In such a search, history plays at least the role of liquidating the seemingly unassailable authority of the scientific establishment and gives one an advantageous perspective.

The sources of discontent among reflective scientists can perhaps be classified into three categories: (1) social, (2) institutional, and (3) internal.

Those who cannot find an acceptable social milieu in which to carry on their scientific research tend to be critical of their own societies, or of the existing social system as a whole. Such was the case of the prewar Japanese generation. Their frustration often emerged early in their scientific careers, if not at the very time when they chose their future careers, that is, precisely when they had reached their most perceptive years. Consequently, they developed an acute sensitivity toward problems related to the social aspect or the social history of science.

Those who are involved in a particular scientific community, yet cannot assimilate themselves to the existing institutional setting, tend to develop critical attitudes toward the current institutions, such as academic hierarchies, university systems, and science policy. In the middle of the 19th century, when scientific work was being established as a profession, young scientists actively participated in the advancement and formation of their own professions. But in the 20th century, now that the professionalization of science has been completed, new recruits are inevitably forced to follow prescribed curricula and must adjust themselves to the standardized behavior, conventionalized values and customary rules of a given community. Professional behavior is no longer a matter of personal choice.

If one fails to conform, he will be labeled "unqualified" and eventually purged from the professional community. The qualification for a historian of science may well be, on the contrary, independence of such conformism. Lacking the necessary critical stance toward current systems, a historian of scientific institutions becomes only a dull, bureaucratic archivist.

Those who embark upon inherited courses toward normal scientific careers only to reach a dead end may, in desperation, seek a way out through historical reflection on their trade. This sort of internal crisis can be recognized only after one has plumbed the esoteric depths of a discipline, and usually does not occur until one reaches the graduate or post-doctoral levels of study. Thus, a critical attitude toward the internal aspects of science often finds its basis in one's own experience of trouble spots. It is doubtful whether future generations of science historians professionally trained at the graduate school level, and lacking experiences of frustration on the research front, can be effectively critical toward the contemporary frontiers of research. They can hardly be expected to overcome psychological barriers which they have never encountered.

Exceptions are most likely to arise from among the recent generation of American historians of science who, despite assured, if not promising, career possibilities, have dropped out of established scientific professions and entered the history of science, presumably because they have, self-consciously or otherwise, traversed one or more of the above paths leading to professional frustration. In scientifically less-developed countries, like Japan in the past, still another element may come into play, namely, dissatisfaction with the pettiness of the research climate in one's environment that prompts one to seek refuge in the historical moments of great discoveries of the past rather than in present reality.

To illustrate the activities of the frustrated, let us review briefly three important, consecutive periods of the history of science in Japan.

Beginnings of the Marxist approach to the history of science

It is commonly recognized among the prewar group that the man and the paper that schocked them in their youthful days[1] was Ogura Kinnosuke and his article "Arithmetic in a class society" written in 1929, or two years earlier than Hessen and his group of Russian historians of science produced a similar shock to science historians at the Second International Congress of the History of Science held in London. Stimulated by G. V. Plekhanov's "Art in class society," Ogura published successively a series of articles on the relation between mathematics and social classes,[2] marshalling historical evidence to demonstrate that even in mathematics, supposedly the most abstract branch of knowledge, class structures are reflected. This thesis created quite a sensation and was widely acclaimed by Marxist-oriented intellectuals in Japan.

When he wrote his first article, Ogura was not too familiar with Marxist docrtine, as he later confessed.[3] In a private conversation with the present writer, he said that his research just happened to coincide with the Marxist approach. Later, after more conscious study of Marxist literature, he extended his research beyond Japan to China, as he felt himself at a serious disadvantage with respect to Western mathematics due to inaccessibility of original source materials. My own view, however, is that he remained throughout his life a liberal democratic thinker and an outspoken critic of authoritarianism, rather than a committed dogmatic communist. This may be one of the sources of his personal attraction and power of persuasion.

His influence was certainly widespread. After the war he was installed as the first president of Minka (Minshu-shugi kagakusha kyōkai [Association of Democratic Scientists], a nationwide movement of progressive scientists), and also as the second president of the History of Science Society of Japan. His most direct influence was exerted on the prewar group of historians of mathematics, such as Kondo Yoitsu and Mita Hiroo, who soon after the war published their works on the social history of mathematics, in a more elaborated and systematic fashion.[4]

Actually, Ogura was a bit older than the prewar generation of Marxist scientists, which included J. D. Bernal, J. G. Crowther, Joseph Needham and Dirk J. Struik—men who were heavily influenced by *Science at the crossroads* (Account of Russian delegates to the Second International Conference on the History of Science). This book played a considerable role in the first widespread movement in Europe and the United States to explore the historical relations of science and society in depth. It was, of course, welcomed in Japan by the prewar group (being put into two different translations), but the way toward development of the social history of science was already paved in Japan by Ogura'a earlier works.

Ogura's interpretation of Japanese science, based on his own historical research, can be summarized as follows:

1. While science in Western Europe played a crucial role in the development of political liberty, Japanese science does not have such a glorious tradition. It was an imported product, and hence naturally imitative and superficial.
2. Since Japanese science has been patronized by feudalistic and bureaucratic governments, scientists have tended to be cowardly, uncreative, and dependent.
3. Governmental institutions monopolize all learning, and hence even natural science retains a strong feudalistic and bureaucratic flavor.
4. The scientific community is controlled by feudalistic academic cliques, who avoid mutual criticism.
5. Scientists are lacking in social consciousness.

These assertions may appear hypercritical of his own tradition, while uncritically laudatory toward modern European science (Ogura was a great admirer of the French Revolution and its scientific institutions). But they can be recognized as strong grounds for resistance against the authoritarian nationalistic bias of the 1930's and early 1940's in Japan. This understanding undergirded the basis for the postwar movement of progressive scientists and was adopted as the official interpretation of the Japanese communists.[5]

Wartime activities

From the late 1920's on, the Communist Party and its sympathizers suffered both police persecution and public condemnation. Their own activities were outlawed, so many scholars concentrated their energies on the analysis of the philosophical implications of Marxist doctrine rather than indulge in concrete realities. The "Yuibutsuron kenkyūkai" (Society for the study of materialism, 1932–1938; cf. III-1) was most active in such philosophical affairs. Ogura was one of the founding members, and Saigusa Hiroto, who after the war succeeded Ogura as president of the History of Science Society, was an original organizer.

Saigusa, after his arrest in 1933, for an offense involving 'dangerous thought,' devoted himself to working on past Japanese thought and to editing a number of Japanese scientific classics. Ostensibly abjuring Marxist belief, he maintained and defended rationalistic thought. History was a safe ground—the older, the safer—to explore even amid repressive measures to control thought. In the late 1930's it became difficult to employ Marxist terms outspokenly, and they were gradually replaced by the vocabulary of science. Under the cover of 'science,' which could not easily be penetrated even by fascist demagogues, the history of science provided for leftist liberals a shelter from the eyes of governmental censorship and from the arms of police thought-control.

The History of Science Society in Japan was founded in the same year that the Pearl Harbor incident occurred. Why so? There was at the time a boom in the history of science among the reading public. One of the main causes of this popularity was the effort to enhance and glorify the scientific achievements of the ancestors in Japan's past, which sought to wipe out the inferiority complex of the Japanese toward Western science and to encourage self-confidence in their cultural heritage.

This boom had a parallel in the enhancement of Aryan scientific contributions in the Nazi ideology, but it does not appear to have been directly connected to the latter. Japanese scientists, even on the extreme right, were not so irrational as to accept Nazi science at face value; it was too ethnocentric to be adopted by non-Aryan races. Adolf Hitler in *Mein Kampf* defined Japanese science as uncreative but culture-supporting. This part was omitted from the prewar Japanese translation and revived only in the postwar version.[6]

According to the editorial notes of the first issue of *Kagakushi kenkyū* (official organ of the History of Science Society), the society was founded in order to correct uncritical popular versions of the history of science current at that time and to demonstrate genuine standards of scholarship in the field. This principle has been well maintained, as testified by the wartime issues of the journal, which was continued up to the end of war despite various difficulties. It was not affected by any nationalistic bias, as the more semi-popular journals were obliged to be.

It is apparent, nonetheless, that founding the society did depend somewhat on taking advantage of popular support. Parallel to that, an extensive series on the history of Japanese science before the Meiji era (*Meiji-zen Nihon kagakushi*) was projected in 1940 to commemorate the 2,600th year in the Japanese chronology. The editing of this series was seriously interrupted by the war, postponing its publication until the postwar period, when it finally appeared in 1954–1968 in twenty-six volumes.[7]

The postwar heyday

Postwar scholarship in Japan has been characterized by the triumph of Marxist doctrines. Primarily because of the persecutions suffered by prewar Marxists, the postwar prestige of Marxist ideology has run extremely high among intellectual circles as a counteraction. Many Marxist-oriented authors writing on science who had been suppressed during the war years now found opportunities for uninhibited expression in in postwar literature. Some examples include Kondo's history of mathematics, Taketani Mituo's three-stage scheme of scientific development, and Hara Mitsuo's advocacy of Engel's dialectics of nature. As a result, Marxist doctrine became established orthodoxy in the period 1945–1950, gaining hegemony especially in the fields of economics and history.[8]

The postwar generation of science historians received the full impact of the postwar version of Marxism in their youth, but their reaction to it differs somewhat from that of the prewar generation. While for the earlier generation Marxism meant a new scientific outlook, a new world-view to be advocated and diffused, it was the academic establishment for the second generation. Once established, it tended, because of attempts to exterminate unorthodox views, to become stereotyped and lose its creativity. Hence, the new generation shares with the old a deep sense of problem but is more inclined to be critical toward the code of the Marxist "Establishment."

For instance, Hirosige Tetu criticized Ogura and his followers on the basis of his own analysis that the postwar leftist movement in science depended too heavily on their assessment of the historical conditions of Japanese society in the late 1920's and early 1930's, and thus failed to face squarely the newly-arisen postwar factors; the earlier group had adhered faithfully to the formula that science is a superstructure determined by its socioeconomic social base, but overlooked the new phase of science established now as a "social institution."[9] Nakamura Teiri made a critical appraisal of the Japanese version of the Lysenko controversy,[10] and Yamada Keiji has tried to find a new perspective on science in China under the Cultural Revolution as a sign of the bankruptcy of modern Western science.[11]

Generally speaking, though, Japanese historians of science are stronger in the externalist than the internalist approach; that is, they stress the social relations of science more than the interaction of scientific ideas themselves. While a great deal of Marxist and externalist literature in Western languages—such as works of J. D. Bernal, J. G. Crowther, D. Struik, Franz Borkenau, and even Sir Eric Ashby—is available in Japanese, no work of Alexander Koyré has yet been fully translated. (So far only an article by him in the *Journal of the history of ideas* has been translated as a part of a collection of essays by many authors.) This externalist characteristic necessarily originates in the historical recollection of the Japanese.

In Western history, historians of science almost unanimously agree that modern science was founded at the time of the Scientific Revolution in the 17th century. But this is true only from the viewpoint of intellectual history. The Scientific Revolution was an intellectual movement of only a handful of scientific intellectuals. Institutionally, modern science was founded, on the other hand, in the 19th century when scientists tried to advance their social status and eventually succeeded in establishing themselves, when their uninterrupted reproduction through a recruitment mechanism based in institutions of higher education was accomplished, and thus when science was fully professionalized. Some people therefore call the 19th century the Second Scientific Revolution. The 19th century gave birth to the present-day behemoth scientific establishment, and hence it is worthwhile to examine the historical origins of the various complex problems existing between contemporary science and society.

When we try to locate the point of departure for modern science in Japanese history, our attention is riveted by the big revolutionary break at the Meiji Reformation in the latter part of the 19th century when Japan embarked on its wholesale importation of Western science and related institutions. Prior to that time there had been, of course, a variety of Japanese scientific endeavors; but it is totally anachronistic to think of Japanese participation in the 17th-century Scientific Revolution. Premodern Japanese had nothing whatsoever to do with the modern Western Scientific Revolution in the internal sense. The Japanese encountered modern science from the very beginning as the developed Western institution of the 19th century. Thus, that century is doubly important to the Japanese since both the internal and the external Scientific Revolutions coincide in the same period of their history. Hence, the Japanese may have a keener sense of the problems of external history of science in recent history than the average Western historian of science who tends to concentrate on the 17th century or an even earlier time.

In the historical recollection of the Japanese, the term 'modern' may summon up the image of American gunboats anchored in Tokyo Bay in 1853; or it may be synonymous with the economic capacity and technological innovation—products of the Industrial Revolution—that supported the military superiority of the West. To the contemporary-minded Japanese, technology has almost equal status with science and therefore the Japanese History of Science Society has a stronger complement of historians of technology than is found in the Society's Western counterparts.

This characteristic was materialized in the recently completed 25-volume *Nihon kagaku-gijutsushi taikei* (Compendium of the history of science and technology in Japan, 1964–70), which deals with Japanese scientific development since the middle of the 19th century. It is, at least in bulk, an astonishing achievement; though judgements as to its quality are in the hands of future historians of science.

Nouvelle vague in the history of science?

Which way the future generation of Japanese historians of science will proceed is not easily predicted. We have seen few fruits from the labors of the "mid-war" generation who came between the prewar and postwar groups, because the devastations of war sapped their youthful energies. We find still less harvest from the new recruits in more recent days, who, thanks to the science and technology boom, have been unwilling to follow such a precarious career as that of the historian of science. Such new recruits as do in fact appear tend to commit themselves to the field only when conditions of normal academic success are assured them. Only very recently, however, there can be seen symptoms of a new generation coming forth—due perhaps to recent university struggles and to pollution problems.

They will face science with entirely new perceptions and pre-conceptions.

For the prewar generation, science was the last gleam of rational thinking to be defended from stormy wartime demagoguery. For the postwar group, science still stood at the center of the democratization chorus. For the new generation, science may appear as a monolithic establishment not easily to be undermined; it is no paradise to be discovered, but only a harsh and inescapable reality. What new picture of science, then, may come out of this "paradise-lost" generation?

In the cold war atmosphere two sciences have prevailed: Japanese science has in actuality consistently and definitely accommodated itself to the American model, contested without much success by adherence in some circles to the socialist version of science. Now that the American model has faltered seriously, are any frustrated young Japanese searching for a third model of science, and if so, will they find a new paradise?

Notes

1. Kondo Yoitsu, one of his pupils, testified to this shock in his article "Ogura Kinnosuke sensei no sūgakushi kenkyū" (Ogura's research in the history of mathematics), *Kagakushi kenkyū*, 1963, p. 17. Cf. also I–1.
2. They are:

   "Kaikyū shakai no sanjutsu" (Arithmetic in a class society), *Shisō*, No. 1, August, 1929—a treatise on arithmetic in the Renaissance period; No. 2, December, 1929—a treatise on Colonial American mathematics.

   "Sanjutsu no shakaisei" (Social character of arithmetic), *Kaizo*, September, 1929—British society and economy in the 16th century as observed in books on arithmetic.

   "Kaikyū shakai no sūgaku" (Mathematics in a class society), *Shisō*, March, 1930—a treatise on the history of mathematics in France.
3. Ogura Kinnosuke, *Ichisūgakusha no kaisō* (Reminiscences of a mathematician), p. 102.
4. See, for instance, Kondo Yoitsu, *Sūgaku shisōshi josetsu* (A his-

tory of mathematical thought; a preliminary study, 1947, Sanishi).

5. Hirosige Tetu, "Ogura Kinnosuke to Nihon kagakushi" (Ogura Kinnosuke and the history of Japanese science), *Kagakushi kenkyū*, 1963: 9–16.
6. *Nihon kagaku-gijustsushi taikei, kokusaihen* (Compendium of the history of Japanese science and technology; international relations), edited by S. Nakayama et al., 1968, p. 333.
7. See Yajima Suketoshi's review in *Japanese studies in the history of science*, No. 7, p. 159 (1968).
8. *Nihon kagaku-gijutsushi taikei, shisōhen* (Compendium of the history of Japanese science and technology, ideology), edited by T. Tsuji (1968), p. 396. Cf. II–5.
9. Hirosige, "Ogura Kinnosuke to Nihon kagakushi", p. 13–15.
10. Nakamura Teiri, *Ruisenko ronsō* (Lysenko controversy), Misuzu, Tokyo, 1967.
11. Yamada Keiji, *Mirai e no toi* (Question for the future), Chikuma, Tokyo, 1968.

Part I

Emergent ideologies of science

1

Arithmetic in a Class Society*

Notes on arithmetic in the European Renaissance

Ogura Kinnosuke

Does mathematics possess a class character? Does it show evidence of having received any social impact? This paper reports on part of my historical studies for dealing with these questions. The subject of this paper is limited to arithmetic, the place to Europe, and the time to the period from the 12th to the 16th century—limitations imposed mainly because of the materials available to me in Japan.

As a result of my studies I have arrived at the conclusion that there existed two contrary types of arithmetic in the Middle Ages: one taught under the auspices of the church, reflecting its participation in the ruling class; and the other practiced by the newly-emerging bourgeoisie, and thus related to commerce and craftsmanship. I do not mean, of course, that purely arithmetical theory or rules differed according to the respective classes—as if $2+3=5$ held true in a monastery but not in a merchant's shop. The two kinds of arithmetic were so different, however, that the differences cannot be explained simply as variations of academic schools and their peculiar interpretations, apart from any social impact. I myself have come to think that the differences derived from the class character of arithmetic, and thus that the two kinds of medieval European arithmetic simply reflected the interests of their respective sponsoring classes.

So as not to burden the reader, I refrain from mentioning all the sources used in my studies, but wish to note the one book to which I am most indebted, D. E. Smith's *Rara arithmetica* (Boston, 1908).

---

* First appeared in *Shisō* (Thought), No. 1, August 1929. Partially translated here as "Notes." See p. 15, ref. 2.

Arithmetic in the medieval church

The problem is made clearer if we discuss first the kind of arithmetic fostered by the medieval church. It is well known that the church, along with lords and knights, constituted the ruling class in the Middle Ages, and that the majority of the ruled were serfs. Moreover, in the feudal system of that period, only the church maintained any substantial program of learning.

Other than the special methods for calculating the annual date for Easter, the church's arithmetic was based mainly on an abridged translation of a text by the Greek mathematician Nicomachus (A.D. 50–110) produced by the 6th-century Roman philosopher Boethius (486–525). The purpose of the arithmetic of Boethius' school was not ordinary calculation as we know it today, but the classification of numbers. For example, even numbers were classified into three categories: "evenly-even numbers," 2^n; "evenly-odd numbers," $2(2n+1)$; and "oddly-even numbers," $2^{n+1}(2m+1)$. Another threefold classification consisted of "perfect numbers," i.e., integers equal to the sum of all their possible divisors, e.g., $6=1+2+3$; "deficient numbers," integers more than the sum of their divisors, e.g., $8>1+2+4$; and "surplus numbers," integers less than the sum of their divisors, e.g., $12<1+2+3+4+6$. The modern reader may be surprised by the Boethius school's complicated classificatory methods and terminology, which often were correlated with religious terms to yield complexities which, it was thought, witnessed to the wisdom of the Creator of the universe.

It may be pointed out, however, that Boethius' arithmetic included no specification of the four arithmetical operations (the rules for addition, subtraction, multiplication and division) which are so essential in daily life. It was largely a non-utilitarian arithmetic of a far lower level than its modern counterpart, or even than the much earlier Euclidean theory of numbers and ratios.

The church-related arithmetic also stressed the mystical nature of numbers, and explained numbers (including measurements) appearing in the scriptures by reference to spiritual matters. Both the numerological and the non-utilitarian nature of arithmetic taught under the church's auspices suggest how well suited it was to the ruling class of that age.

The introduction of Indian arithmetic by the bourgeoisie

Against the background of emergent handicraft industries, the rise of free commercial cities, and the interaction of cultures through the Crusades, there began in the 12th century the period of translation of Arabic mathematics, which had, in turn, been shaped by the earlier Greek and Indian traditions of mathematics.

One of the outstanding geniuses to come from the new merchant class was Leonardo Fibonacci (1174?–1250), who was born in Pisa, one of Italy's great commercial centers. As a child he attended a Muslim school in a city on the coast of northern Africa where his father was engaged in cxport-import business. As a young man Fibonacci traveled widely in Egypt, Syria, Greece, Sicily and southern France, before returning to his home town of Pisa. Once settled, he completed an Indian-Arabic style textbook on arithmetic (*Liber abaci;* 1202), the first to approximate the standards of modern arithmetic.

Fibonacci's book was the first to use the Indian way of writing and reading figures; it contained the four rules of arithmetic, fractions (decimals were not yet invented), and explanations of commodity exchange, pricing, partnerships and other commercial uses of arithmetic. Indeed, Fibonacci was the first to introduce commercial terms into arithmetic, and he gave more pages to practical applications than to theory. Among these applications are a number of problems of Indian and Arabic origin, a few of Greek and Roman pedigree, and a couple thought to have originated in China—all well systematized and genealogically noted. There are also several problems which compute interest at the rate of 20 per cent, and an explanation of the term 'capital' as well. Here can be clearly seen the influence of early capitalism.

[Ogura then discusses various details of arithmetic in the

13th, 14th, 15th and 16th centuries; these sections are omitted, as the purpose here is to show the concern and way of thinking of this pioneer work, rather than historical facts.]

Conclusions

The contrast between the two kinds of arithmetic of the ecclesiastical and mercantile classes appears quite evident. Churchly arithmetic based on Boethius' theory of numbers did not include methods of calculation using Indian symbols, it had virtually no relation to daily life, and stressed the occult significance of numbers. In contrast to this, the arithmetic of the bourgeoisie consisted mainly of calculations using Indian symbols, and stressed commercial applications. Arithmetic in the universities was an intermediate combination of Boethius' theory of numbers and calculations using Indian symbols, but omitted commercial uses.

Not least impressive is the fact that the Indian symbols used from the 13th century by merchants, while neglected by intellectuals, became quite popular among ordinary townsmen by the 16th century. Boethius' number theory was soon eclipsed, as various textbooks with higher-level theory and applications were written by talented men of the period when commercial arithmetic flourished.

The reader may well ask, "Why did no arithmetic arise among the peasants?" The answer is that in the process of the collapse of feudalism only the bourgeoisie were liberated; behind their successes was hidden the poverty that still cursed peasant life. The children of merchants and craftsmen in cities and towns had their schools, and sons of the nobility went to cathedral schools. For peasant children, there were almost no chances for schooling. How could they produce and develop their own arithmetic?

As Plekhanov showed in the case of art, so in mathematics the strongest impact came from the class that ruled society during the Renaissance, the church where and when it dominated, and the bourgeoisie in the free cities where they were masters. Thus, arithmetic did not merely reflect the living patterns of a ruling class; it had a definite class character. The arithmetical rules, discovered by human genius, were themselves, of course, independent of any social relations. But the knowledge preparatory to discovery represented the accumulated insights of past periods, and the direction of the discoverer's interest was greatly conditioned by class background.

2

Methodological Approaches in the Development of the Meson Theory of Yukawa in Japan*

Takctani Mituo

In this article we shall give a brief account of the methodological approaches and the circumstances surrounding Yukawa's proposal which formed the background of the development of the meson theory in Japan. The remarkable discovery of the meson theory by Yukawa might be thought comparable to the sudden appearance of a beautiful flower in sterile ground. The real state of affairs was just the reverse. The ground had already been made fertile for natural growth.

Theoretical physics in Japan before Yukawa

A nuclear atom model was proposed in 1903 by Nagaoka Hantarō,[1] about the same time that the able English physicist J. J. Thomson[2] suggested an atom model without a nucleus. Nagaoka held that an atom should consist of a massive positively-charged nucleus with a number of electrons revolving around it in a ring. The ring of electrons should vibrate and create many spectral lines describing the patterns of vibrations.

This atomic model was called Saturnian and was somewhat different from a solar-system model of an atom. Using Maxwell's theory of Saturn's rings, Nagaoka calculated the vibration frequencies of the revolving electrons. In 1911 Rutherford[3] proved the validity of a nuclear atom model, thus giving support to Nagaoka's model of an atom with a nucleus and refuting Thomson's non-nuclear model. The main difference between Rutherford and Nagaoka was their estimation of the number of electrons contained in an atom. While both Naga-

* Adapted from an article in Supplement of the *Progress of theoretical physics*, No. 50, 1971, translated from an essay in the book *Shinri no ba ni tachite* (In the course of our study) by H. Yukawa, S. Sakata and M. Taketani (Mainichi Shimbun, Tokyo, 1951).

oka and Thomson speculated about thousands of electrons, Rutherford estimated that the electrons in an atom were few. N. Bohr[4] later showed that the lightest atom, hydrogen, has only one electron. This caused a serious difficulty, for if an atom has only one electron, then the line spectra with several kinds of vibration frequencies are unaccounted for on the basis of classical electrodynamics. Also, such an electron revolving around a nucleus could not continue in a fixed orbit. Atomic physics thus confronted a decisive difficulty; and in fact, with classical electrodynamics we cannot understand an atom. Bohr in 1913 solved this problem by introducing a new quantum theory[4]—ten years after Nagaoka first suggested a detailed model of a nuclear atom.

That a Japanese physicist achieved this successful result in the initial stage of atomic theory gave a great impetus to contemporary Japanese physicists. Unfortunately Nagaoka had no successor. One of the main reasons may be that a feudalistic atmosphere then prevailed in academic circles and new activities were suppressed. When Nagaoka presented his atomic model before a meeting of the physical society in Japan prestigious professors charged that such a theory was metaphysics, not science. Nagaoka was so discouraged that he turned to the study of magnetism.

The second achievement in this field was brought about by Ishiwara Jun.[5] He was the first of the theoretical physicists in Japan with a character different from Nagaoka. Ishiwara was attracted to Einstein and his work on relativity. In 1911 Ishiwara visited Germany and studied under Albert Einstein and Arnold Sommerfeld, became an acquaintance of Max von Laue, and returned to Japan in 1915 to work on relativity.[6] In 1911 he wrote a paper[5] discussing the validity of the quantum theory of the photon, and in 1915 introduced a generalized coordinate system[7] in which he formulated the quantum condition which is basic for the quantum theory of the atom. His formula served as a correct theoretical context for atomic physics. This work was published earlier than analogous work by Sommerfeld.[8] His theory played an important role when quantum theory was constructed for various general cases, thus serving as a bridge leading to the new quantum theory. Although he contributed much to the development of theoretical physics in Japan, he was expelled from Tohoku University because of his personal love affairs. This too happened in the feudalistic atmosphere of an imperial university. Afterward he engaged in the popularization and reviewing of scientific work. He edited the Japanese scientific journal *Kagaku* (Science) of Iwanami Publishing Company. His contribution to the promotion of science in Japan is undeniable. While some laborers in the field of science are noted primarily for their own research achievements, we must not forget the organizer type who creates an atmosphere that stimulates others to fruitful work. Ishiwara should be appreciated as one who cultivated fertile land for the later development of theoretical physics in Japan. Many currently productive theoretical physicists in Japan were influenced by him.

Nishina Yoshio then appeared as a theoretical physicist whose work was accepted internationally. Returning to Japan after studying theoretical physics under Bohr in Copenhagen, he founded the Nishina Laboratory in the Institute of Physical and Chemical Research (Rikagaku Kenkyūsho) in 1931 in Tokyo, and began research on nuclear physics and cosmic rays. He is known for the Klein-Nishina formula[9] published in 1929. Klein and Nishina used the Dirac equation to study the scattering of gamma rays by an electron, and the resultant formula was shown to agree with experiment. This was an important proof of the Dirac theory, and Nishina, along with Nagaoka and Ishiwara, came to be noticed by the international community.

My intellectual pilgrimage

Soon after I advanced to the third class of Kyoto University the infamous "Kyoto University incident" occurred. Minister of Education Hatoyama dismissed Takikawa Yukitoki, a professor of law, without the customary approval of a formal meeting of the law faculty, provoking an atmosphere of resistance. In discussing this incident with friends I took the position that the Minister of Education was entirely wrong; indeed, his action was analogous to the Nazi suppression of knowledge which had just appeared in Europe. Under Nazism many valuable books of international reputation were burned, and many famous physicists and other scientists idolized by us all were expelled from their positions. We felt that similarly absurd events were being precipitated by the military leaders of Japan. Earlier in 1931, Japanese military forces had invaded the Manchu territory of China (the "Manchurian incident") as fascism escalated in our country.

Around 1934, when I graduated from Kyoto University, radical progress was being made in the physics of elementary particles. Chadwick's discovery of the neutron in 1932 had led to clarification of the structure of the nucleus. Various difficulties in understanding the nucleus were solved by the new model, which includes the neutron as one of its constituents. Then the positron was discovered by Carl Anderson in cosmic rays and by Joliot-Curies in radioactive nuclei. Formidable difficulties in relation to electron theories had been resolved before the discovery of the positron. Strongly influenced by these facts, I had now to reconstruct my opinions from an entirely new standpoint regarding the logic of science.

At that time Neo-Kantism and Machism dominated scientific thinking. It was thought physics discovers laws by manipulating empirical facts, and only sense impressions are real; all the rest is constructed by the subject who arranges experienced facts. Moreover, it was maintained that the essence of science lies in function, and therefore, in the course of scientific development, substance will be absorbed in function.

The thinking of Ishiwara Jun and Tanabe Hajime, Kyoto University's leading philosopher, was more or less along these lines. If applied to physics, such thinking made it nonsense to consider the internal structure of the nucleus—with a diameter of only about 10^{-2} cm.—or its structural elements. It is not the role of science to arrange mathematically the various phenomena which concern the nucleus.

The actual ways in which physics developed were completely different. So long as a correct model of the nucleus was lacking, all knowledge relating to it was in confusion. However, when a new substance, the neutron, was discovered and a correct model of the nucleus was available, the confusion was resolved. Likewise, before the positron was discovered, the electron theory of Dirac had been burdened with inexplicable difficulties. Once a new substance, the positron, was discovered, all questions were solved at a stroke, and related knowledge was systematized. With Dirac's theory one could deal with various phenomena from a general viewpoint. I observed parallels with the processes of recognition of the heliocentric system and the construction of atomic theory. Furthermore, the theory of beta decay was similarly developed. Bohr suggested that laws of the conservation of energy and of other quantities might not apply to the interior of the nucleus, a suggestion that led to the breakdown of rational theory of the nucleus. But in 1931 Wolfgang Pauli proposed a new particle, the neutrino, to accompany the electron in beta decay. To this proposal Fermi applied the methods of radiation theory and succeeded in formulating a complete theory of beta decay. In this case, the introduction of a new substance, the neutrino, rescued a discredited theory and made it possible to treat theoretically the heretofore intractable phenomena of beta decay. And for me the value of innovative thinking about the methodology of science was convincingly demonstrated.

In April 1934 I had occasion to visit Sakata Shōichi, who had just procured a position at Osaka University under Yukawa, and was informed about these new developments in discussions with the two of them. This chance alone afforded me a clue on how to attack anew two problems in particular which troubled me at that time. First, the prevailing views on science were incapable of producing new developments in physics; it was necessary to explore new approaches to the methodology and logic needed to promote studies of nuclei and cosmic rays. Second, the development of quantum mechanics had a profound impact on contemporary philosophy, yet the philosophers' interpretations of quantum mechanics seemed superficial. I sensed the need to develop a philosophy capable of treating properly the new insights into matter by appropriating the new logic of quantum mechanics. Thus, I began studying the mathematical foundation of quantum mechanics.

Sakata, who was well acquainted with natural philosophy, kindly helped me in my investigations. I had already reviewed various schools of philosophy and explored their application to my problems, but realized that I had to go directly to the original sources, not interpretations by contemporary Japanese philosophers. As I came into intimate contact with original works by the great philosophers, I felt that my long-standing problems were solved. This is the real method of attacking science, I thought, for I have managed to master each step of my philosophical pilgrimage through many complications and surprises.[10]

Yukawa's theoretical breakthrough

This brings us to the point of discussing the process of construction of the meson theory by Yukawa. It should be recalled that the problems of nuclear force and beta decay were then dealt with mainly from the phenomenological side. On the other hand, the field theory was developed and enthusiastically used in many sophisticated trials. Only a few people attacked the problems of beta decay and nuclear force from a fundamental standpoint.

Fermi[11] treated beta decay on the basis of the field theory by using the neutrino hypothesis. Tamm and Iwanenko,[12] in turn, calculated nuclear force by using Fermi's theory of beta decay. They gave the short-range nuclear potential in terms of a model in which nuclear force is derived by the interchange of the electron and anti-neutrino between the proton and neutron. The resultant strength, however, was smaller than that observed by a factor of 10^{-15}. Again, the problem was accentuated. Bohr[13] favored the view that quantum mechanics cannot be applied to the electron in the nucleus, and also rejected Fermi's theory. He was especially unhappy with the hypothesis of the neutrino, which had not yet been observed. Bohr's inference that energy is not conserved in the nuclear region was reinforced by the beta-decay theory of Beck and De Sitter (1933) using the mechanism of pair-creation. At the time it was generally considered that the system of laws itself was wrong. There were few who thought that a new substance should be introduced. Instead, they warned against speculative introduction of a new substance. Such is the work of dilettantes, not serious scientists. Professional physicists must rely on confirmed experimental data and overcome difficulties with highly specialized mathematical techniques. This was the current trend among physicists. It is interesting to note that an analogous trend is now evident in work on elementary particles.

Yukawa started from the theory of Heisenberg,[14] who assumed that protons and neutrons are the constituents of nuclei. He first imagined that a neutron (or proton) will change into a proton (or neutron) by emitting an electron (or positron), producing nuclear force. He could not, however, obtain a reasonable result from this assumption based on quantum mechanics. Heisenberg's model of nuclear force had some defects in this respect. Yukawa then learned of Fermi's theory and was impressed by its far-reaching implications. Sakata also accepted Fermi's theory from his standpoint of natural philosophy, by which the mysterious principle of Bohr had been criticized and the reasonable principle of Fermi was welcomed. In Tokyo, Sakata had already studied with Tomonaga[15] the problems of pair-creation by nuclei before the gamma ray is emitted. Yukawa and Sakata[16] utilized Fermi's theory to discover the mechanism of electron capture by the nucleus, a process later proved by the experiment of Alvarez.[17] Yukawa had a very cooperative and productive co-worker in Sakata.

Yukawa became quiet excited when he studied the calculations by Tamm and Iwanenko of estimated nuclear force based on the assumption that an exchange of electron—anti-neutrino pair between the proton and neutron leads to a nuclear potential of short range. Faced with the difficulty that the result obtained by Tamm and Iwanenko was too small to give the strong nuclear force resulting from the weak beta-decay interactions, Yukawa postulated a new substance—a revolutionary decision.

According to Yukawa,[18] an entirely new field must be introduced in order to derive a strong nuclear force. Furthermore, the quantum of such a field must have a mass big enough to yield a short-range potential. (The mass of the quantum was estimated to be $\sim 200\ m_e$, m_e being the mass of the electron). To this field he gave the name "*U*-field," or "heavy quantum." The *U*-quantum should have a plus or minus electric charge in order to produce the nuclear force between the proton and the neutron. With regard to beta decay, the *U*-field acts as intermediary: first, the proton (or neutron) produces U^+ (or U^-), then U^+ (or U^-) decays into $e^+ + \nu$ (or $e^- + \tilde{\nu}$). Just as the light quantum plays an important role in the domain of atoms and molecules, resulting in the quantum theory, so this new heavy quantum plays a role in the nuclear domain.

The above is an outline of Yukawa's revolutionary idea. His report, completed in October 1934, was presented before the annual meeting of the Physico-Mathematical Society of Japan on November 17, and was published in the Society's journal the following year.

Of course, Yukawa's work was not yet recognized by anyone. This was partly because Japanese science was not well known abroad, but mainly because such a quantum had never been observed. No one was willing to accept a theory based upon the existence of an unknown particle. In Japan no one except Nishina and Tomonaga accepted the Yukawa theory. Those of us who were close to him welcomed it, of course, because his theory provided a fresh example for testing the validity of our methodology. The theory itself remained vague and in need of fuller development, but its aims were, we thought, the grandest.

In retrospect we can agree that the introduction of a new particle was natural and no great risk. Perhaps now it is clear that no special methodology was necessary. However, it is important to recall the atmosphere at that time. In choosing among many possibilities a proper methodology is needed to get started on the way to truth. Once the results prove correct and natural, the methodology used has less importance. But when an unknown domain is still to be explored, confirming the accuracy of one's methodology is essential. People who were engaged only in interpretation, and had never wandered in the dark of the unknown, could not appreciate this necessity.

Collaborative follow-up

In the mid-1930's many young scientific workers in Kyoto, representing various fields, organized a research group outside the university and engaged in lively discusssion about resistance movements. They began publishing a journal called *Sekai bunka* (World culture) in January 1935. I joined this group and took part in discussions with many people from different cultural fields, and to the journal contributed articles on contemporary physics and on methodology. Meanwhile, Japanese militarism escalated and the whole nation was subjected to strict imperialistic controls, including thought control by a special police force. In Europe the Nazi pressures grew ever stronger and an anti-fascist movement was organized to resist them. In France, the popular front won a victory and formed a government to carry out progressive reforms.

Fascism in Japan would destroy culture and research activity, if not stopped, and our learning would collapse as it had in Germany. *Sekai bunka* not only published papers on new research activities but also enthusiastically discussed cultural movements of the popular front against fascism in Europe. The efforts of *Sekai bunka* were meant to support resistance to Japanese imperialism—which at that very time was making preparations for invasion of the Chinese continent. Nakai Shoichi insisted that if the invasion were permitted, the Japanese intellectual class would be shamed.

My endeavors concentrated on exploring physics, particularly to establish a methodology useful in promoting actual research, not just interpretations of achievements already made, as was current among philosophers. My papers on the logical structure of quantum mechanics (1936) and on the three-stage theory of the development of physics were published in *Sekai bunka*. Sakata has spoken admiringly of these papers.

First consideration went to the possibilities of the postulated meson, in order to develop Yukawa's theory. We wondered whether nuclear force could be derived in every respect from the meson theory. Yukawa and Sakata calculated the nuclear potential from the fourth order perturbation. With a charged scalar meson the results are not satisfactory. In this calculation the quantized scalar meson field formulated by Pauli and Weisskopf[20] was first used, and it appeared that the sign of the nuclear potential given by the first paper of Yukawa was of the opposite sign. In view of this result, they suggested that a neutral meson is necessary in addition to the charged mesons. Meantime I was paying attention to the anomalous magnetic moment of the neucleon. By means of the correspondence principle, I was led to suggest that the spin of a meson is 1, if the magnetic moment was to be derived from the meson field around the neucleon. Just before that time Yukawa and Sakata had formulated the generalized Dirac equations for a spin greater than 1. A third paper on the meson theory was prepared by Yukawa, Sakata and Taketani,[21] in which the neutral meson was naturally introduced as a partner of the charged mesons in the symmetric meson theory.

In the spring of 1937 Bohr visited Japan and gave some very exciting lectures on quantum mechanics and the role of observation. Audiences were impressed by his enthusiastic and sincere attitude. Yukawa and Nishina met him in Kyoto, and talked about the meson theory, to which Bohr was not attracted. We were discouraged that he asked Yukawa, "Why do you want to create such a new particle?" However, before Bohr returned to his native country, we were informed from the United States that a new charged particle having a mass of $\sim 200\ m_e$ (m_e being the mass of the electron) had been found in the cosmic rays by Anderson and Neddermeyer,[22] and also by Street and Stevenson. In the Nishina Laboratory of the

Institute of Physical and Chemical Research in Tokyo, Takeuchi examined the photographs obtained by the cloud chamber and found tracks of the new particle. A report was published in the September issue of *Kagaku* by Nishina, Takeuchi and Ichimiya,[23] and we thought, "The predicted particle is at last discovered." The difficulties of nuclear theory were solved by the discovery of a new particle, not by modification of the system of laws such as the electron theory or quantum electrodynamics, as had been expected by the majority of physicists.

In 1937 Yukawa[24] published a short note in the journal *Proceedings of the Physico-Mathematical Society of Japan* titled "On a possible interpretation of the penetrating component of the cosmic ray." In this note he suggested that the new particle may correspond to that predicted by himself in 1935. We were very encouraged and became absorbed in further development of the meson theory.

Throughout these thrusts into the dark it was the three-stage theory of materialistic dialectics that guided our research and fortified our resolution to overcome difficulties. We maintained the following viewpoint of our methodology: In contrast to current opinion that difficulties of nuclear theory lie in some deficiency in quantum mechanics itself, the problems of the meson theory should be attacked within the framework of quantum mechanics by using the correspondence principle. Our method focussed on substance. That is, we thought that nuclear physics was at the stage where its main problems were the investigation of what the constituent substances might be, and what might be the properties of these substances. We felt that we were still far from the stage at which quantum mechanics itself should be fundamentally modified.

When we completed formulation of the spin-1 meson, we were informed that Proca[25] had already worked out the same equation as ours, and that the quantization was made in the U.S.S.R. Against the background of confusion in nuclear theory, we were able to analyze step by step the reasons for the tremendous difficulties, and to properly reorder the problem. This was the merit of our methodological attack based on the three-stage theory of materialistic dialectics. With idealistic Machism we might have become lost or even led to an overall denial of the theory.

Sakata was steady and had a keen analytical ability. Yukawa was endowed with great creativity in complicated matters. When the new formulation had been developed and the academic, mathematical formulae were assembled, Yukawa deduced the physical implications. Sakata was a genius at mobilizing people to collaborate; no one felt hostile toward him. He could always get a cooperative atmosphere by speaking frankly.

By the end of February, after our paper was completed, we read in the January issue of *Nature* that similar papers had been published in the *Proceedings of the Royal Society* by H. J. Bhabha,[26] N. Kemmer,[27] and also by H. Fröhlich and Wo Heitler.[28]

Scientists seeking refuge

In the summer of 1937 military forces of Japan started to invade the Chinese continent. It seemed that we had worried about Japanese imperialism and resisted it in vain, though we had not been able to do much. The Special High Police suddenly clamped down and the newspapers and radio broadcasts generated a militaristic atmosphere overnight. It was with very gloomy feelings that I continued studying the development of the meson theory.

In the autumn the active members of the publishing group of *Sekai bunka*, with whom I was collaborating, were suddenly arrested by the police. *Sekai bunka* was an entirely legitimate journal; its circulation had never been prohibited nor its publishers warned by the police. Before this crisis we felt that we must not diminish our determination to fight against fascism; but now it was obvious that we could not continue to publish the journal. I did, however, discuss counterplans with Kobayashi Keinosuke, a biologist who had joined the *Sekai bunka* activities following the Kyoto University incident.

During the period of my intensive work on the meson theory,

I sensed that pursuit by the Special High Police was imminent and, to escape this danger, I left Kyoto in the winter of 1938 and took up lodging with an acquaintance in Kobe. We referred to such changes of lodging then as 'seeking refuge'. It happened that I could meet with Yukawa and Sakata more frequently because they lived near Kobe, and this made it easier to coordinate our research.

Toward the end of June 1938 a girl friend of Kobayashi brought a letter from him to my lodging, informing me that the remaining members of *Sekai bunka* had all been arrested on June 26. Some journalists reported that I too was being sought by the police. For the time being I had eluded them by abandoning Kyoto for Kobe. Had I been arrested in the autumn of 1937, I could not have collaborated in the work on the meson theory. A fugitive with a premonition of imminent arrest, I could waste no time but had to do as much as possible.

Three days after I had once again changed my Kobe lodging, some plainclothesmen forced themselves into my apartment while I was asleep and arrested me. It was six o'clock in the morning of September 13. I was detained at the Fukiai police station, and later was taken to Kyoto and held at the Uzumasa police station. There I met Shinmura Takeshi, who had just been interrogated by the police. He informed me of the whole situation. After a month I was taken to the Kawabata police station, which was responsible for investigations related to Kyoto University. There the police began examining my case. In a memorandum I had to make a detailed account of my activities in connection with *Sekai bunka*, explaining the change and development of my thought in relation to a number of events. The unlawful acts for which I was held were: my analyses of quantum mechanics, my analyses of the development of nuclear physics, and my methodological approach to the meson theory—in short, my research activities on natural dialectics. I was forced to state that I had participated, through my research, in the cultural movement of the popular front under instructions of the Comintern, thus helping to promote the Communist Party in Japan.

At that time I was an assistant (without salary) at Osaka University and also had the same position in Kyoto University. Around the end of October my brother came to visit me at the police station and told me that the universities would not be happy if I continued to occupy my positions. Sakata, on the other hand, tried to conceal my arrest from the universities, and even tried to obtain my release. When the universities learned of my arrest, they wanted my resignation instead of making efforts for my release. It bothered them that I occupied even unsalaried assistantships, so I resigned. In February the investigation by the Special High Police ended. I was allowed to study in the investigation room before the inquiry was started by the public prosecutor. I was allowed to communicate with Sakata about problems in physics and to read photocopies of the latest research papers which he sent me. I was not permitted to take the papers to my cell, though I dared to do so, hiding them under the blanket. I sent a letter to Sakata, saying that we must consider separately the life-time of the meson and the life-time of nuclear beta decay. In the investigation room I calculated exactly the magnitude of the coupling constant of the strong interaction for the nuclear force, which led to a better result for the life-time of the meson. About one week later I received a typed paper by Yukawa and Sakata showing that the same result was reached by them. It pleased me that our results, obtained quite independently, coincided.

In April 1939, when a prosecutor examined my case, I showed him a number of our papers which had already been referred to by scholars in foreign countries. He seemed to have some appreciation of our work in physics. After several days he summoned Yukawa and released me to his custody under suspension of prosecution.

References

1. *Philosophical magazine* (London), Vol. 7 (1904), p. 415.
2. *Ibid.*, Vol. 9 (1906), p. 769.
3. *Ibid.*, Vol. 25 (1913), p. 10.
4. *Ibid.*, Vol. 26 (1913), p. 1024; Vol. 27 (1914), p. 703.

5. *Proceedings of the Tokyo Mathematico-Physical Society*, No. 6 (1911), pp. 81, 164, 201; No. 8 (1915), p. 326.
6. *Ibid.*, No. 8 (1915), p. 173.
7. *Ibid.*, p. 524.
8. *Atombau und Specktrallinien* (1924).
9. *Zeitschrift für Physik*, Vol. 52 (1929), p. 853.
10. *Progress of theoretical physics*, *Supplement*, No. 50 (1971), p. 27.
11. *Zeitschrift für Physik*, Vol. 88 (1934), p. 161.
12. *Nature*, Vol. 133 (1934), p. 981.
13. Faraday Lecture in the *Journal of the Chemical Society* (1932), p. 383.
14. *Zeitschrift für Physik*, Vol. 77 (1932), p. 1.
15. *Scientific papers of the Institute of Physical and Chemical Research* (Tokyo), *Supplement*, Vol. 24 (1934), No. 17, p. 1.
16. *Proceedings of the Physico-Mathematical Society of Japan*, No. 17 (1935), p. 467; No. 18 (1936), p. 128; *Physical review*, Vol. 51 (1938), p. 677.
17. *Physical review*, Vol. 54 (1938), p. 486.
18. *Proceedings of the Physico-Mathematical Society of Japan*, No. 17 (1935), p. 48.
19. *Ibid.*, No. 19 (1937), p. 1084.
20. *Helvetica Physica Acta*, Vol. 7 (1934), p. 709.
21. *Proceedings of the Physico-Mathematical Society of Japan*, No. 20 (1938), p. 319.
22. *Physical review*, Vol. 51 (1937), p. 884.
23. *Ibid.*, No. 52 (1937), p. 1198.
24. *Proceedings of the Physico-Mathematical Society of Japan*, No. 19 (1937), p. 712.
25. *Journal de physique et le radium* (Paris), Vol. 7 (1936), p. 347; Vol. 9 (1938), p. 61; *Journal de physique* (Paris), Vol. 7 (1936), p. 347.
26. *Proceedings of the Royal Society* (London), Vol. 166 (1938), p. 501.
27. *Ibid.*, p. 127.
28. *Ibid.*, p. 154.

3

On Concepts of Technology*

Hoshino Yoshirō

The principle of practice

The formation of our thinking on technology can best be introduced by referring to several affirmations of Taketani Mituo. "It is beyond question," he says, "that technology is a concept of *praxis*. . . . It is necessary to examine practice from the inside to see what makes it possible, and how it is carried out. . . . Hegel's famous observation, quoted by Lenin and others, that 'freedom is an insight into inevitability' has guided my thinking on this, providing me with two basic viewpoints concerning technological practice.

"The first is that human practice, and particularly productive practice, conforms to objective laws. There is no human practice which can disregard objective laws.

"The second is that technology and skill are two different things. Without distinguishing between them, it is impossible to grasp correctly the historical development of technology and to cope with present technological difficulties.

"This separateness of technology and skill is precisely the reason for stressing technology's conformity with 'objective' laws, not 'natural' laws. While both have their basis in natural laws, technology is objective and natural, whereas skill is subjective and natural."[1]

What is meant here by productive practice is, of course, labor itself. In terms of Marxian economics, labor constitutes one aspect of the production process of capital. In chapter 5 of Volume I of *Das Kapital*, Marx comments:

> Labor is, in the first place, a process between man and nature, in which man affects, regulates and controls his

* Published in Hoshino Yoshirō, ed., *Taketani Mituo chōsaku shū, 4, kagaku to gijutsu* (Collected writings of Taketani Mituo, Vol. 4, science and technology; Tokyo, Keisō Shobō, 1969), pp. 356–370.

> metabolic interaction with nature through his own action. He opposes himself to nature's materials as one of nature's own forces. He sets in motion the natural forces of his body —his arms and legs, his head and hands—in order to make natural materials available in forms useful to his own life.*

If labor were in essence no more than this, it would be no different from such animal activities as eating feed and building nests. There would be no essential difference between men and animals. The principle that makes human labor what it is, must be considered. Or, to repeat Taketani's assertion, "It is necessary to examine practice from the inside to see what makes it possible, and how it is carried out."

After discussing human labor as a natural force, Marx continues:

> We presuppose labor in a form that stamps it as exclusively human. . . . At the end of every labor process, we get a result that already existed in the imagination of the laborer at its commencement. He not only affects a change of form in the natural material on which he works, but he also realizes in it a conscious purpose of his own that governs his modus operandi, and to which he must subordinate his will.

These two points, made by Marx with regard to the fundamental nature of labor, merit our attention. The first is that man confronts nature as a natural force himself. The second is that, beyond this, he realizes within nature his own purpose, which acts as a law in determining the way in which he operates. For this reason, labor is first and foremost governed by natural laws. Labor cannot function if natural laws are disregarded. Furthermore, these natural laws are made manifest

* The author' quotation is from the Japanese translation of the German edition of *Das Kapital,* of which chapter 5 corresponds to chapter 7 in the classical English edition (prefaced by Engels; cf. *Complete works of Marx and Engels,* Vol. XXIII, p. 234). Here and below the quotations from Marx were translated directly from the German edition, adapting the translation of the English edition where appropriate. It should be noted that the sentences in the English edition often differ from the corresponding ones in the German edition, not only philologically but also, to an extent not negligible, logically. -Ed.

in labor through the intermediary of this purpose. Thus, they are not mere laws of nature, but ones which accord with the purpose of man.

The essence of that which makes labor what it is, then, is the conscious application in work of natural laws suitable for some purpose. Animals merely behave in accordance with natural laws and are, therefore, unable to free themselves from them. They have no choice but to continue being animals forever. Man, however, has conscious goals. He grasps objectively and subjectively the natural laws suitable to his purpose, and consciously applies them to practice. Because of this fact, man's behavior, while depending on natural laws, is also free from them. Within the natural realm, man is able to build civilized society and to bring about his own continual change. This is one thing Hegel meant by 'freedom is an insight into inevitability'.

Taketani defines technology as "the conscious application of objective laws in productive human practice."[2] What he calls "objective laws" can be interpreted to mean the objective natural laws suitable to human purposes. Most probably the reason why he specified "productive human practice," and not simply labor, was to place the concept of *praxis* within an epistemological context in order to show the importance of the theory of technology for epistemology, and thus that the essential relationship between cognition and practice exists not merely in the area of production but extends also to politics, economics, scientific research, and all other areas of human practice.

Furthermore, the reason why he says objective laws, and not natural laws, is to make clear the distinction between the two modes of man's purposive dealing with natural laws. When a steelworker makes a judgment, from the color of the molten mass he sees in the furnace, that the pig iron has completely turned to steel, his awareness of the natural laws involved in steel production depends on his visual perception, that is, on his subjective grasp of natural laws. What is generally called intuition relies on this kind of subjective grasp.

Taketani calls it skill, which he clearly distinguishes, in principle, from objective technology. Says he: "Technology is objective, whereas skill is something subjective, psychological and personal, acquired through experience. Because technology is objective, it is systematic and social in nature and can be passed on from person to person in the form of knowledge. This means that technology is gradually enriched by the transmission and accumulation of knowledge as part of human society's development."[3]

On the relationship between technology and skill, he says further: "Labor functions as a union of technology and skill. That is, labor is realized when a given technology is accompanied by the particular skill which necessarily goes with it. The proper function of technology is, however, always to transform subjective personal skill into objective technology. This does not mean that skill is no longer needed; far from it. New technology requires new skills, which are, in turn, further transformed into new technology. Thus, both technology and skill evolve in a dialectical relationship. Nevertheless, as skills are transformed into technology, there is generally a tremendous rise in production capacity and an improvement in product quality."[4]

Thus, Taketani's definition of technology is directly related to the basic meaning of labor, since, clearly, there can be no other logical definition of the essence of technology. The essential characteristics of technology and skill appear as purposive functions of such factors in the labor process as labor and its means, objects, materials and products, that is, as the suitable technical abilities of laborers and technicians, and appropriate efficiency of the tools, materials, and products of labor.[5]

The logic of the relationship between science and technology

When technology is thus understood, the essential relationship between science and technology becomes self-evident. Science is concerned with universal natural laws. It penetrates deeply into nature in an attempt to establish empirical laws, based on experiments and observations, as universal laws. Technology concerns itself, however, not simply with natural laws in general or with regularity in nature, but with specific natural laws which are amenable to its own specific goals. While science is not content with empirical laws but seeks to confirm universal natural laws, for technology the most important thing is acquisition of a product suited to its own purpose. This requires, not the elevation of empirical laws to universal status, but the conscious application of purposively selected natural laws.

Science and technology develop through close interaction, though each has its own special logic of development. The development of science is often required by technology, but from the standpoint of science this demand is, so to speak, an external and fortuitous factor. Following the logic inherent in nature, science searches for ever more universal and integrated laws. Novel phenomena that appear in technical operations may, therefore, become interesting subjects for scientific research, even if their elucidation is not technologically important. One good example is the contemporary field of solid state physics, which developed from efforts in quantum mechanics to shed light on such phenomena as the ductility, elasticity, and the destruction of various materials. Although solid state physics found its object of study in the area of actual production, there was no particular or direct demand for its development from this source. Furthermore, present-day solid state physics has not provided any immediate help in improving the quality of materials or in developing new materials. It has developed more or less in accordance with its own logic, yet with stimulus from production technology.

Technology also serves as the basis for progress in science by providing new experimental apparatus. It is hardly necessary to mention how much the astronomical telescope, the vacuum pump, various instruments of measurement, the electron microscope, the large particle accelerator, the plasma generator and other apparatus have contributed to scientific progress.

Therefore, demands put on science by technology constitute no more than one of the forms of the technological basis for scientific development.

Taketani says of this relationship between science and technology: "Demands on science from technology indicate the direct connection between the two. The advance of science is governed by three factors: technology, the structure of nature itself, and modes of thought. Though technology sometimes makes direct demands on science, the connection between the two is not always a direct one. Science and technology are but links in a larger chain, and often interact with each other indirectly."[6] In Taketani's scheme scientific knowledge, while also conditioned by technology and modes of thought, has its own logic of development by which it advances from the phenomenological stage through the substantial stage (i.e., recognition of substances and their structure), to the essentialist stage (i.e., recognition of essential laws) based on the intrinsic logic of nature.

Demands from technology and negations of accepted ideas in scientific theory by technological practice, then, help to further scientific progress. However, were science to neglect its own deep and broad quest based on the logic of nature, and only respond to technological demands as they arose, it would be impossible for it to make genuine progress. To be sure, Marconi's successful communication by transatlantic wireless led scientists to the discovery of the ionosphere. But science did not stop there; it went on to the study of cosmic ray physics, and further to astrophysics and elementary particle theory. Eventually such self-development of science in turn brings about an enormous development in technology. It is well-known, for instance, that nuclear technology issued from the discovery of nuclear fission. This discovery, however, was totally unrelated to any idea of the application of nuclear energy, but was, rather, the result of investigation of the essential nature of transuranic elements as an important part of systematic research on nuclear disintegration through neutron bombardment.

Contradictions between ends and means

While technology obviously depends upon the regularity of nature, the scientist's pursuit of natural laws is for technology an external and fortuitous factor. What directly concerns technology is purposive application of specific natural laws to achieve certain goals. The general laws of nature are, of course, extremely important as a support system for technology, but in some cases it is also necessary to ignore some accepted ideas in scientific theory, in favor of the necessary consideration of the goal itself. A goal or end is the use value yet to be realized, and therefore exists in one's mind or plans. What the technician concerns himself with is basically determined not merely by natural laws but also his own purpose. For this reason, he must not only consider suitable natural laws but also ask the meaning to human society of the goal itself. In philosophical terms, the search for natural laws is a problem of reality, while the meaning of the goal to human society is a question of value judgments. In this sense, technology involves the integration of existence and value, and thus technological theory seeks to integrate ontology and axiology. In this respect, too, Taketani gets at the core of the problem when he sees the essence of technology in the definition of freedom as "an insight into inevitability."

Naturally, the technician must also ask himself the actual meaning to human society of a product as a realized goal. But that is not all—he must also ask himself the meaning of the labor process to human labor. Not only does man contribute to present-day society by producing things with his labor, but the possibility of contributing still more to future society arises from the fact that through labor he grows and acquires the ability to realize goals of an even higher order. This is the motive force behind the development of human society. Like the means and objects of labor, labor itself is a means of attaining the goal in the labor process, but unlike them, it is an independent human entity which enhances its function through the labor process. It is this growth factor of human labor which, along with realization of goals, constitutes the fundamental significance of the labor process to human society.

Any given goal can be attained by one of the several ways in which suitable natural laws can be applied. An extremely important criterion for deciding which way to adopt is the extent to which it allows for the maturation of the labor force through the labor process. The next important criterion for deciding which of the possible purposive applications of natural laws to adopt is the desirability of acquiring as much product as possible with as little labor input as possible. We mean here not only the present labor input but also the labor expended to produce the tools and materials of labor. In the labor process, labor creates more use value than is necessary for its own maintenance. This peculiar function of labor explains why it is the basic source for the development of productive forces, and the basic condition for the progress of human society as well.

If labor were not able to produce enough for its own maintenance, man would have starved and the human race would have perished. Furthermore, if labor had produced only enough use value to maintain itself and not enough for its growth, the human race would have, if not perished, stagnated forever in a primitive state for lack of developed productive capacity. It is precisely the fact that labor is able to produce at least enough to maintain itself and its growth that has made it possible for human society to develop to where it has today. In any case, the greatest importance must be attached to human growth through the labor process when considering which of the possible purposive applications of natural laws to adopt.

The reason why human labor can grow through the labor process is that new and higher goals are imposed on this process, with the result that contradictions arise between new goals and existing means (i.e., the labor force, and the tools, objects, and materials of labor). To deal with these contradictions, man tries to determine the most efficient set of natural laws needed to obtain the most product with the least labor. A thorough examination of all possibly relevant natural laws is made by relying on his intuition, on empirical laws, and on advanced theories of natural science. Appropriate natural laws are chosen and expressed in the way labor and its means and materials are mobilized to establish a new labor process.

We have said that contradictions arise between ends and means when a higher goal than before is set. But what gives rise to the higher goal? First of all, a contradiction that emerges between the end and means in a given labor process can stimulate the setting of a new and higher goal directly related to that particular labor process. In other words, making every possible improvement in the various means (labor force, tools, materials, etc.) can be made a new goal within the existing labor process. When the end is achieved and a product materializes, that product then becomes a new means for another labor process. Thus, the contradiction between ends and means in one labor process gives rise to a new contradiction between ends and means in another, resulting in an endless chain reaction.

This chain reaction in the labor process constitutes the basis for the formation of a system in which technology in any one area is closely interlinked with that in others. Although the idea of the gas turbine presented itself with the invention of the steam turbine at the end of the 19th century, realization of that idea had to await development of heat-resisting steel as the material for the turbine vane. With use of the jet propulsion of the gas turbine, the speed of airplanes rapidly approached the speed of sound. To break the sound barrier, however, it was necessaary to adopt thin wings and then sweepback and triangular wings, which meant radical changes in fuselage design as well. Electronic control systems also became necessary for such high speeds. Earlier, technological research in aviation had been about equally divided between the structure of the plane and its propulsion system. With the advent of the jet, however, electronic systems came to command fully as much research resources as did structure and propulsion together. With the introduction of jet-propulsion systems in passenger aircraft as well, pressure has been mounting for an overhaul of land transportation systems in an effort to eliminate the speed imbalance between land and air transport, and to offer an alternative to the noise of jet aircraft. Because technology in each area is closely interlinked with that of others in the whole system of technology, technological innovation in any one area leads to an immediate chain reaction of ends-means contradictions and thus to innovations throughout the entire system. This is the first ground for the constant setting of new goals.

The second case is one in which the goal or end is not the production factors in one or another labor process, but rather consumer goods which have a bearing on such basic human needs as food, shelter and clothing. Consumption is not merely consumption but also reproduction of labor. Or, conversely, mobilization of labor and its means and materials for production is also a form of self-consumption.

Consumption and production mutually consummate each other. Clothing which has been produced first becomes clothing only when it is worn. Clothing not worn is really not clothing at all. Only when the fruits of production have value in use can they be considered fruits of production, and consequently only then can production be considered production. Also, only when labor recovers its vitality and heightens its physical and mental powers through consumption does it become human labor in production. Says Marx: "Granted that eating, drinking, giving birth, and so on are truly human functions; but if they are separated from other areas of human activity and abstracted in such a way as to make them final, unique, ultimate goals, they become not human but animal functions." Consumption becomes human consumption only when it sustains human labor for production.

One might say that production and consumption simultaneously are separate yet the same. Just as there is demand for improvement in the quality of individual units of labor in the labor process in order to realize new goals, in consumption there is demand for more advanced consumption. Thus, contradictions between ends and means arise in the consumption process as well. For example, aspirations for higher living standards introduce contradictions that are overcome by improving the food, shelter, clothing, and (through cultural improvement) the consumer. If the contradictions grow to such proportions as to defy solution within the scope of the existing productive capacity, they become contradictions no longer internal to consumption but between the production and consumption systems. Raising the goals of production gives rise to contradictions within the labor process which, if overcome, permit increased production, and thus improved consumption. This, in turn, makes possible improvement in the quality of labor and, thus, solution of the contradictions in the labor process. So it is that contradictions between ends and means in the labor process give rise to new goals, whose solution makes possible heightened consumption, improvement of labor quality, solution of the contradictions between ends and means, and finally solution of the contradiction between production and consumption.

The third ground for setting new goals is change in nature itself—increasing production and consumption results in the exhaustion of natural resources. At the end of the 19th century a shortage of Chile (cubic) saltpeter imposed on the chemical industry the new task of fixing atmospheric nitrogen, and at present the increasing depletion of operative coal seams makes it necessary to set such new goals as hydraulic coal mining and underground gasification of coal. The exhaustion of water resources that has accompanied increased production has made desalination of sea water an urgent task of the present day.

Fourthly, man as a natural force increases his own numbers through his reproductive activity, causing new goals to be set to meet the consequent increase in consumption. While such reproductive activity is in itself a form of consumption, it can also be thought of as the production of new labor. At the same time that it imposes new goals because of increased consumption, it also furnishes new means whereby contradictions between ends and means can be overcome.

Here we have discussed fundamental mechanisms unique

to technological development and distinguished from those of scientific development. Our argument has unfolded logically from Taketani's view of the essential character of technology, how an abstract proposition gradually leads to concrete considerations. Logical propositions correctly abstracted and centered on the core of the matter in question can, because of their abstraction, be extremely rich in content.

Notes

1. *Kagaku to gijutsu*, Vol. 4, (Keisō Shobō, 1969), pp. 136–137.
2. *Kagaku to gijutsu*, p. 139.
3. *Ibid.*, p. 137.
4. *Ibid.*, p. 138.
5. Cf. the author's paper "Restatement of the theory of technology (1)," in *Ritsumeikan economics*, Vol. III, No. 6 for details.
6. *Kagaku to gijutsu*, p. 84.

4

Toward a Truly Free Labor Force*

A review of Nakaoka Tetsurō's *The future of man and labor*

Shizume Yasuo

The author is quite honest, I think, in pointing out that no one today is capable of presenting, with intellectual conviction, a clear and convincing vision of the future of Japanese society, much less of world society. Yet a number of valuable suggestions can be drawn from this book.

Quoting Rousseau's *Emile*, Nakaoka says that before the Industrial Revolution—in Japan, prior to the Meiji period—the craftsman, "should he be subjected to persecution, could, no matter where he was, pack his belongings in a bag and leave his workshop, relying on his own two hands." And Nakaoka concludes his book with these words: "Perhaps I am only dreaming of a modern version of the freedom 'to pack one's bag and leave, relying on one's own two hands'."

This reviewer's dream is more heroic. For the past ten years I have advocated a philosophy of peaceful coexistence. In my opinion, progressive forces in Japan should have anticipated such international developments as West Germany's diplomatic stance and the Italian and Canadian policies for recognizing China, long before Prime Minister Tanaka took advantage of the new political openness by recognizing China last year.

The vital point in my outlook stems from the policy that the best way toward a truly bright future for mankind is to promote the "freedom to leave (even one's country), relying on his own two hands," for ordinary people everywhere—the United States, the Soviet Union, Japan—and not just those in divided countries like Germany and Korea. This kind of freedom existed in the past for a few outstanding artists, artisans and thinkers, who were often forced to use it; for ex-

* Published in the journal *Manejimento* (Management; Jan. 1971), pp. 134–135.

ample, China's Confucius in the 6th century B.C., Anaxagoras and the young Plato of the ancient Greek World, Avicenna and Ibn Khaldun of the medieval Islamic world, and Leonardo da Vinci of the Renaissance period. Even today such freedom is still assured only for outstanding artists, scientists and engineers outside the communist countries.

To diffuse this freedom to ordinary men and women, it is necessary that, for every people with a common language and shared customs and living standards, there be two optional types of national states. One of these necessary types is a non-Stalinist communist government, and the other a reformed capitalist government. Only through such freedom can the existing governments overcome internal corruption and rigidity, much less move toward the discovery and creation of a third and truly new form of democratic society for the future of mankind.

In his book, Nakaoka says that Rousseau, Pestalozzi and Marx all considered 'labor' as indispensable to a truly human life and self-realization (self-development), and that what they meant by 'labor' was the work of the craftsman. Nakaoka also takes the craftsman as the ideal prototype of free human labor, involving the total personality. Moreover, in today's highly industrialized society, it is only the labor of high-level personnel in large enterprises and research institutes, he says, that retains the same flavor as the craftsman's labor. In his words, "the greater portion of their work is still based on their personal training and experience."

But the author also asserts, "Although the modern elite craftsman's labor is undoubtedly enjoyable, the greater part of his enjoyment derives from his advantageous position to use the power of the organization for high achievement and at the same time expand his own ability. He completely confuses his own talent with the organizational power at his disposal (sometimes negatively manifested in the embezzling and swindling by key executives of large corporations and banks, recently so much in the press). Consequently, the world of the businessman, and even of the scientific researcher, increasingly resembles the medieval church in terms of organizational enormity and of the power and scope of control over others. And the mentality of top administrators in such organizations seems also a reversion to the medieval world."

The author warns that such trends are no longer confined to the capitalist system alone. While capitalism's 'separation of management and ownership' (manager and capitalist) has brought no improvement in the situation whatsoever, in the communist countries as well the people are controlled by only a small number of administrators—though in the name of all the workers. Even the triple combination of workers, technicians, and administrators promoted in China during the Great Cultural Revolution, while highly evaluated for its achievements thus far, likewise embodies the danger of deteriorating into a medieval-style mechanism.

Thus, while sympathetic toward the ideal future society of which Marx dreamed—ownership of the means of production by the workers and cooperative labor of the masses—the author expresses doubt about the possibility of realizing this ideal.

Will the future of man and labor turn out to be, after all, as some sociologists predict, only "a small elite who get busier and busier, while the masses whose time is not well utilized become more and more numerous?"

The actual situation in Japan is still a long way off from a society in which the masses have more free time they than know what to do with. The vast majority of Japanese are driven like cogs in the computerized and mechanized systems to which they must adapt themselves physically and mentally. Equally important, the developed socio-economic mechanisms that control these systems make interpersonal relations both in and apart from work increasingly indirect; and because of the interposition of machinery between man and man, ordinary persons are increasingly under the illusion that they are controlled by machines when they are actually controlled by other people.

Even this book seems colored to some extent by this illusion, which has already been overcome by the concepts of technology developed by Taketani Mituo and his followers. According to them, technology consists not in the systems of machines constituting the means of production, but rather in man's conscious application of certain objective laws of nature to productive purposes, or in the working to carry out that application (cf. pp. 39–42, this volume). This book does, however, elucidate very well that aspect of any highly industrialized country in which not only ordinary people, but also top-level administrators in business and government are in fact controlled by their own social mechanisms, rather than being in control of them.

To break out of this unhappy condition, Nakaoka says that labor unions should no longer concentrate their efforts on winning higher wages and shorter hours, as in the past, but in restoring their work to that which "cultivates their basic human abilities." To do so, they must gain more discretionary powers over actual work processes and greater rights of participation in management. To acquire these powers and rights, ordinary factory and office workers must join in larger citizens' movements outside their own places of work (especially joint-struggles with local residents in specific communities), while also strengthening cooperation with the technical and administrative personnel within the enterprise where they work. On these points I agree with the author wholeheartedly.

Part II
Exploratory research concerns

1

Reflections on the History of Science in Japan*

Ōya Shin'ichi

The purposes of science history

The histories of modern physics and mathematics can be helpful when considering what directions to promote in these disciplines. Histories of biology and medicine as academic disciplines have roots farther back in the past, yet have connections with present research in these fields. But the history of Japanese science during the Edo period (1603–1868) does not have the same kind of relations with contemporary research as does the history of science in medieval Europe. The study of Edo science is more like studying the science of ancient Egypt or Babylonia. Edo science differs from that of those ancient cultures only in that it was the science of our ancestors; that is the main reason for us to study it.

According to the preface of Dannemann's great work on the history of natural science, he wrote it to regenerate the self-respect of youth battered down by defeat after World War I. In relating the history of world science in the past he would clarify the achievements made by German scientists and thereby stir the spirits of German youth at that time, and embolden their hearts to diligence in science and in all their life—that was the purpose of his book.

The rapid rise of research in science history in China following World War II derived from a similar motivation. The research undertaken was mainly on sciences peculiar to China; that is to say, the sciences of Chinese antiquity. The researchers regarded their purposes in doing Chinese science history in this way: the Chinese people are not lacking in scientific ability; indeed they are superior to other peoples. This much is clear from the scientific achievements of the Chinese in the

* Published in *Kagakushi kenkyū* (Journal of the history of science, Japan), Vol. 9, No. 95 (1970), pp. 129–132.

past. But from China's middle ages Chinese science stagnated due to defects in the social system. As Westerners intruded into Chinese society, their disdain for Chinese science gravely affected the Chinese people, who came to think poorly of their own original capabilities in science. This was a mistake, as Chinese scientific abilities are in fact higher than those of other peoples. Today, when the defects of the social system are being corrected, the Chinese can, through their own efforts, raise the level of their science and make their own contributions to the science of the world. This sort of emphasis has been dominant among recent Chinese science historians.

The Chinese science historians go beyond Dannemann in making the interrelations of science and society the basis of their thought. This is possibly due to the influence of Japanese science historians like Mikami Yoshio and Ogura Kinnosuke, as many of their writings were translated into Chinese prior to World War II. But taking encouragement of the people as the purpose of the history of science is common to Dannemann and the Chinese alike. A similar trend can be seen in postwar Soviet research in science history.

To make science serve such ends is not necessarily proper. Science history, as an independent discipline, has its own purposes. To grasp accurately the conditions of scientific development, and to reflect upon the causes and laws of that development, are its proper purposes. Considering the realities of our actual engagement in studies of science history, however, it is hardly possible to deny completely any relations to the age we live in. Ultimately we have to face a choice between learning for learning's sake and learning for the sake of mankind. This problem is made especially acute if we attempt to treat science as if it had virtually no relation to society, as was the case with science in the Edo era.

Edo science certainly should be studied for its intrinsic value; but if contemporary implications are ignored, there is the danger that it will be nothing more than a hobby. That is permissible if learning is for the sake of learning alone. But if that danger is to be avoided, it is imperative that we reflect from time to time on the contemporary significance of our research on the history of science in Japan.

In asserting this I do not have in mind doing research, in the manner of Dannemann, simply to display the superiority of Japanese science. Except for a few minor examples, it would be, after all, quite difficult to establish any superiority of Japanese science. From very ancient times our country has been underdeveloped with respect to science. Scientific ability, however, is another matter. Despite a number of inhibiting conditions, our ancestors demonstrated considerable ability. It is their weak points, though, more than their strengths, which are useful in helping us determine goals for contemporary science. The defects of our ancestors are our defects too; it is these which to a large extent hamper the development of science in our nation today. Thus, to evaluate these weaknesses and act on the lessons learned is to promote scientific development along the right lines in our time, I believe.

The history of Japanese science is an independent academic discipline, and essentially should be studied as such; although we should avoid doing it in complete isolation from relations to the present age.

Proper objects of research in the history of science

The research objects of science history can be broadly divided into the following:

(a) the development of science as such;

(b) persons who promoted science; and

(c) social conditions that aided or inhibited scientific development.

There are, within science itself, developmental elements. Consequently, it is possible to study the development of science as such quite apart from any persons or social conditions which affected its development. In the past all science history in our country was of this kind.

Especially Edo science, or that which preceded it, because they have little relation to modern science of the present day, were made the objects of research just as they were, not as part of historical development in science. Much as the papyri of ancient Egyptian mathematics or the tablets of ancient Babylonian mathematics were translated and studied for their intrinsic value, so in our country the texts of the special mode of Japanese mathematics called *wasan* (fl. 17th century on) were converted into modern notations, and that passed for "research in the history of *wasan*" (though not as research on *wasan* as such). Even this constitutes one basic task of science history, but to stop with only that is problematic. Without asking what methods were used to solve *wasan's* problems but, instead, just solving them by modern methods, does not constitute suitable research on *wasan*.

I have made an example of *wasan*, which is unusually distinct from modern science, but the point applies to all sciences. Though Edo science is made the object of research, the researchers themselves have been trained in modern science, and the low level of Edo science is very easily understood on modern terms. For the history of science, though, the proper research method is to place oneself in the historical situation of the scholar at that time and, starting where he started, to discover or reconstruct his line of thought and how his successors further developed it, adhering closely to the source materials left by them. Of course, the researcher's own interpretations creep in, but that is more or less inevitable for the historian.

In the long run, science always depends on the continuous accumulation of achievements, and the developmental character within science derives largely from this fact. Consequently, to isolate some portion of science and study it as a stationary, static thing usually distorts our results. Even when researching some scientific achievement of a specific period or person, it is necessary to consider in what way it was dependent upon cumulative achievements. In other words, the purpose of research is not reconstruction of the intrinsic history of science alone, but also of the social conditions of a particular period, or of the labors of a particular scientist, not separated from all else in that period but as part of a larger historical flow.

Given the cumulative nature of science and the developmental elements within it, whatever promotes or inhibits its development is found in the social conditions of a given period. Whether the overall volume of scientific achievement increases rapidly or only very sluggishly is determined by social circumstances. Accordingly, considerations of scientific development must always take into account the state of society at that time.

The study of natural history in the Edo period is a case in point. It was called *honzōgaku* (pharmacognosy) then, and its early development had two purposes: one, to aid in the understanding of the Chinese classics, and two, to identify medicinal drugs. Early in the Edo period, as a result of Tokugawa policy to encourage learning, the printing and study of Chinese classics flourished; but for adequate understanding of the classics a knowledge of flora and fauna referred to in the classics was essential. Especially in books like the *Book of songs* (Chinese, *Shi-ching;* Japanese, *Shikyō*) comprehension of the content was virtually impossible without knowledge of the animals and plants mentioned. Likewise, it was equally important for the Japanese to identify counterparts to the flora and fauna in their own country. Gradually such studies were extended also to the classics of Japanese tradition. In time there appeared a variety of large reference works which collected and catalogued all records of flora and fauna in both Chinese and Japanese classics.

Work in the initial stage was limited to cataloguing and classifying, but in the process of determining exact counterparts—or whether those found in China actually existed in Japan or not—the work of collecting and examining specimens became indispensable. It was in such circumstances that botanical and zoological research in Japan were begun. The same process served simultaneously to fulfill research needs in pharmacology.

Japan imported drugs from China in great quantities in the Edo period, putting such a strain on national resources that the shogunate made a special effort to increase domestic production of materia medica. Botanical and zoological surveys were initiated, with a corps of specialists itinerating the provinces to collect and examine various specimens. *Honzōgaku* thus became directly related to practical realities, and reached its peak in the time of the eighth shogun, Yoshimune, who also promoted domestic production. Among the distinctive products of each locality there are many for which production started in this era. *Honzōgaku* began thus to develop as a form of productivity research.

Late in the 18th century people in Japan became extravagant in their desire for novel things, and this affected *honzōgaku* as well. Exotic plants were cultivated and unusual animals bred. Even special teahouses for viewing the beauties of flowers and birds (*kachō-chaya*) appeared. The wide variety of plants and animals became objects of botanical and zoological research. Among leisured persons there emerged many amateur practitioners of *honzōgaku*, which became widely diffused as it was also removed from purely utilitarian concerns. In such circumstances European natural history was transmitted to Japan, and just as natural history in Japan was on the threshold of becoming a pure science in its own right, the Meiji Restoration occurred.

As the development of Japan's natural history was intimately connected with the social conditions and demands of the time, so were the other sciences more or less related to society in their respective developments. Mikami Yoshio was the first in our country to stress these interrelations between scientific developments and social conditions, followed by Ogura Kinnosuke. Japanese initiatives in this approach to the history of science, then, occurred quite early. However, while recognizing close relations between society and science, Mikami placed even greater emphasis on the relation of the scientific enterprise to the character of the Japanese people, and hence on the persons who engaged in the cause of science. It was Ogura who most strongly advocated attention to relations between science and society.

Both China and Japan have, in modern times, introduced Western science; but the mode of reception has clearly differed.

In China translations were usually done by teams of two persons; specifically, a foreigner fluent in spoken Chinese would read aloud the content of a text, while a Chinese colleague recorded it in written Chinese. This reading-recording system had evolved much earlier in the process of translating Buddhist canons into Chinese. The Matteo Ricci group used the same system in the 16th century to translate scientific treatises, as did the Hobson group in the 19th century. For roughly eighteen centuries the method of translating foreign texts into Chinese went unchanged. Using this system, the translations were from the outset rendered into fluent and accurate Chinese. On the other hand, no cumulative skills in translating were developed. The Chinese were consistently poor in independent translations.

The Japanese did not adopt this system of collaboration with foreigners. If available, they welcomed opportunities to receive instruction in foreign languages; but for translations it was most common for some Japanese, by himself or with a few other Japanese, to do the actual translating. Consequently, the initial results were quite clumsy with many distortions of the original text. But there was cumulative improvement in skills, and translations gradually became fluent and accurate.

These processes were not confined to translation. To understand the "heaven-origin algebra"* of China, that is, algebra done on the counting board, the Japanese proceeded to study Chinese mathematical texts by themselves. Their struggles to master this complex method provided the motivation for Seki Takakazu's eventual invention of written notations to replace the counting board, giving a unique character to Japanese mathematics in the Edo era.

A strong trend toward self-reliance, then, was clearly the strength of the Japanese mode of adopting foreign culture; yet, at the same time, there was also, in the long tradition of assimilating foreign culture, a weakness in accepting

* Traditionally, though mistakenly, rendered "celestial-element algebra." "Heaven" and "origin" are the coefficient powers of unknowns. -Ed.

it uncritically. Late in the 16th century when Christian missionaries first introduced the notion of the earth's sphericity, almost everyone believed it as reported. Hayashi Razan refuted this idea on the basis of the ancient Chinese dictum that heaven is round and the earth is square, but he was an exception. Some interpret Razan's case to mean that many resisted the theory of a spherical earth; but that is incorrect, for there is no other example than Razan himself. At that time it was popular in Japan to paint oval-shaped world maps on folding screens; the maps showed the course of Magellan's circumnavigation of the earth, indicating that people in general accepted the sphericity of the earth with no resistance at all.

Furthermore, from the time of introduction of the heliocentric theory by the Dutch in the middle of the Edo era, the situation was the same. Only a few Buddhist priests, such as Entsu, resisted this theory, on the basis of Buddhist doctrines; but scholars of National Learning (Kokugaku) like Hirata Atsutane accepted the heliocentric theory as transmitted, as did apparently the general public.

Since the theories of a spherical earth and heliocentricity are correct, acceptance of them was no mistake. But people of that age believed these theories not because they were right but because they were transmitted from foreign countries—that is the problem. Today, too, theories from abroad are accepted and believed unconditionally. There is a certain tendency to be critical toward theories proposed by fellow Japanese, and this is not new. Historically, though, the uncritical stance assumed toward things foreign seems to be one reason for the scarcity of creative research initiated by Japanese.

The character of the Japanese people was shaped over a long time by social conditions, though, and subservience to foreign ideas was not an innate trait of our race from the very beginning. It is reasonable to think that as society changes, so the people change. This being true, research on scientific developments of a given period should always take into account its relations to the social conditions that formed the basis of that development in that time. Such an approach allows also for treatment of the characteristics of the persons involved in those scientific developments and influenced by the same set of social conditions.

Trends in research on the history of science in Japan

Genuine science history in Japan began in the Meiji era (1868–1912), the earliest instance being the 1877 publication of *Nihon yōgaku nenpyō* (Chronology of Western learning in Japan) by Otsuki Nyoden. Not limited to science, it nonetheless belongs to this field because of the intimate relations to science of *yōgaku*, that is, Western learning introduced into Japan during the latter part of the Edo period. Expanded and revised by the author, this work was republished in 1927 as *Shinsen yōgaku nenpyō* (Newly-compiled chronology of Western learning); and because of continued demand, was further expanded by Satō Eishichi and reissued in 1965 as *Nihon yōgaku hennenshi* (Annals of Western learning in Japan). From the outset and through subsequent revisions it was more than a mere chronology, as each entry was well annotated.

Shirai Kōtarō's *Nihon hakubutsugaku nenpyō* (Chronology of natural history in Japan), with detailed explanations, appeared in 1890, followed by the author's own expanded version in 1908, and finally a further expanded and revised edition, based on the deceased author's research papers, in 1934.

Dai Nippon sūgaku-shi (A history of mathematics in Japan), by Endō Toshisada, was published in 1896. Originally in chronological form, the author was at work on an expanded version at his death; Mikami Yoshio used his research notes to produce an enlarged edition in 1918, on which, in turn, Hirayama Akira based an expanded "definitive" edition in 1960.

Fujikawa Yu's *Nihon igaku-shi* (History of Japanese medicine; 1904), and his own "definitive" revision published in 1941, complete the list of major works in the history of science during the Meiji period. In contrast to other works noted above, the initial and revised editions of Fujikawa's work differed little in content. Moreover, it was a more systematic causal description of scientific development than were the earlier chronologies and annals. Fujikawa's work on medicine and Endō's on mathematics were the two most important Meiji era productions in science history.

Nothing comparable to the above four major Meiji works appeared in the Taishō era (1912–1926). Only one volume, *Nihon no kagakkai* (Japan's world of science; 1917), can be counted. A general outline of Edo and Meiji science based largely on the above four works, it was well received by those concerned with science history because no other single, handy general survey existed. Published by the Association for Japanese Culture (Dai Nippon Bunmei Kyōkai), the actual compilation and editing were done by Densei Sukeshige. It was the most widely consulted work of its kind until around 1935.

Early in the Showa era (1926–) Mikami Yoshio and Ogura Kinnosuke began writing essays on the history of mathematics, which altered the focus of science history by taking mathematical developments as examples of cultural evolution, especially in terms of interrelations with social history. Sawada Goichi published his *Nihon sūgaku-shi kōwa* (Discourses on the history of Japanese mathematics) in 1928; sections on the Edo era were basically educational in purpose, but this volume introduced many pre-Edo materials not yet treated.

The mid-1930's were especially important to science history. Commemorating its first anniversary, the Tokyo Science Museum in 1934 published *Edo jidai no kagaku* (Science in the Edo period), a volume containing many photographs (of items exhibited in the museum) of great value to science historians; though the text contains many errors and can be used only with great caution. Kanda Shigeru edited two source books (*Nihon tenmon shiryō sōran*, 1934; *Nihon tenmon shiryō*, 1935) of documents on pre-Edo solar and lunar eclipses, and data on planets and comets; though not histories of astronomy as such, despite the titles, these volumes provided access to valuable source materials.

Sekiba Fujihiko's *Sei-igaku tōzen shiwa* (Accounts of the transmission of Western medicine; 1933) in three volumes performed a similar service for medicine by compiling records of the Edo era. Fujikawa's work on the history of Japanese medicine proved so influential that no one attempted systematic research in this field after his death in 1940; though many articles on medical history were penned, as researchers and enthusiasts were most numerous in this field. Essays on *wasan* (indigenous Japanese mathematics, flourished from the 17th century) begun around 1930 were issued in a first collection by Ogura Kinnosuke in 1935, prompting renewed efforts on this specialty of the Edo era. For example, Hayashi Tsuruichi published a two-volume study of *wasan* in 1937.

Apart from the labors of Mikami and Ogura, research in the 1930's centered on the source materials of science history. Comprehensive histories of science began to appear only during the early part of World War II. The History of Science Society of Japan was founded in 1941, the year that the Imperial Academy (Teikoku Gakushi-in) began compilation of its series *Meiji-zen Nihon kagakushi* (History of pre-Meiji Japanese science). Work on this series ceased temporarily when the war ended, but was later resumed under the renamed Japan Academy (Nihon Gakushi-in), and many volumes on mathematics, astronomy, physics, chemistry, biology, pharamacology and medicine were published from 1954 on.

Most typical of the war years was publication of many short volumes. Tominari Kimahei's brief *Nihon kagaku shiyō* (A short history of Japanese science; 1939) on Edo science and the introduction of Western science was one of the earliest and the first comprehensive treatment of science history. Ogura Kinnosuke published a collection of radio talks on *wasan*, *Nihon no sūgaku* (Japanese mathematics; 1940), that developed his ideas on relations between science and society in easily-understood form. A more specialized treatise was *Wasan shisō no tokushitsu* (Distinctive features of *wasan* thought; 1941) by Hosoi Sō, but like the previous two works by Tominari and Ogura, it too treated science against the background of social conditions. Other wartime works included Araki Toshima's *Nihon rekigaku-*

shi gaisetsu (Historical outline of Japanese calendrics; 1943) and Katō Hirasaemon's *Wasan no kenkyū: gyōretsu-shiki oyobi enri* (Research on *wasan:* determinants and principles of circle mensuration; 1944).

The Japan Academy's volumes on pre-Meiji science introduced research in certain fields for the first time; for example, biology (1960), chemistry (1961), and physics (1964). Ueno Masuzō, who edited the Academy's volume on biology, an excellent work based on primary sources, also produced the first comprehensive history of zoology. Except for treatises on mathematics, and particularly *wasan*, few histories of specific sciences appeared other than the Academy's volumes, although the publishing of numerous monographs on the histories of medicine, natural history and mathematics in specific provinces and localities constituted a unique postwar phenomenon. Few of the writers of these local histories went on, however, to do more general research.

A far more significant postwar trend was the publication of several general histories of science. Yoshida Mitsukuni's *Nihon kagakushi* (History of Japanese science; 1955) is unexcelled in its detailed treatment of science prior to the Edo period. A small but methodologically excellent work is the *Kindai kagaku shisō no keifu* (The genealogy of modern scientific thought; 1964) by Horiuchi Gōji. Three scholars—Sugimoto Isao, Satō Shōsuke, and Nakayama Shigeru—co-authored *Kagakushi* (History of science; 1965) for inclusion in a larger series on Japanese history. That the three works mentioned here were, in fact, all part of more general historical series indicates the place finally won for science history within Japanese historical studies. Finally, Asahi Shinbun-sha published a large volume, *Nihon kagaku gijutsu shi* (History of science and technology in Japan; 1962), that consists of short descriptive essays by different specialists on a wide variety of subjects. The most general postwar survey is the *Meiji-zen Nihon kagakushi sōsetsu, nenpyō* (Outline of pre-Meiji Japanese science, with chronology) published in 1968 by the Japan Association for the Advancement of Science (Nihon Gakujutsu Shinkōkai).

Directions for research in the history of science in Japan
Following publications in the past of a number of authoritative works, general and specialized, one would expect work to have continued on a variety of specific themes. Instead, Japanese research on the history of science has declined drastically. One reason for this is that few universities offer teaching positions for science historians. The few scholars active in this field do their research as an avocation while teaching some specific science. The number of specialists in science history is incredibly small, and this condition is unlikely to change until changes are made in the university system itself.

Meantime, we should endeavor to increase the corps of professional science historians. One inhibiting factor here, apart from job scarcity, is the difficulty in identifying suitable and attractive research topics. To overcome this difficulty there are two possible avenues: one is use of a reliable history of science; the other is exposure to some primary sources. In actuality, the first way is hardly exciting with no knowledge of primary sources, while the second does not always make sense with no general overview in mind. In all likelihood the best method to arouse an abiding interest is a combination of both avenues. One helpful road to acquiring competence is to take an authoritative text in science history and attempt to improve its reliability by locating and correcting any errors. Such can usually be done only by turning to primary sources. This, in turn, not only aids general comprehension of both the general work and the original source, but also can serve as a seedbed for the conceptualizing of attractive and worthwhile research themes.

Moreover, I think this process can best be done in small research groups. Mutual stimulation in a research group not only strengthens research motivation but critical thinking as well, and can lead to the discovery of issues not previously considered. Some recent works, such as Nakayama Shigeru's assessment of Asada Goryu's derivation of formula for variation of tropical year-lengths, or the studies of early Edo mathematics by Kanda Shigeru, Shimodaira Kazuo, and Noguchi

Taisuke, are cases in point. Also, many previous treatises failed to include sufficient reference to primary sources used, and exploration to supply these deficiencies affords a useful entry into science history. Finally, many important source materials are not yet available in printed form. A most fruitful mode of research is the thorough analysis and subsequent printing of one of the sources, with a scholarly commentary appended. Producing a solid commentary inevitably forces a researcher to comprehend both related texts and relevant social data.

The recovery of primary sources is so vital for the history of science that it is mandatory for annotated editions to be given the widest possible circulation. Naturally it is best if a responsible publishing house will undertake the printing and distribution of these materials. This is not always possible; but even so, a research group of, say, ten members can achieve great things if each member will see to the reproduction, by mimeographing or photo-offset printing, of one manuscript apiece and share his copies with at least other members of the group. For a scholar to have his own printed copy of ten important primary sources is no small thing. Persons working with private means can legitimately place a suitable price on distributed copies of their work, especially if made available to a larger audience.

There are many yet unexplored topics in the history of Japanese science, including problems in the Edo period. As each researcher delves more deeply into his chosen subject, he will develop a keener sense of his own vocation as a researcher. His own thinking processes will, of course, color the results of his labor; but this too can have a salutory effect on others, as did Sugimoto Isao's analysis of the impact of Confucian thought on the rationale of "practical learning" (*jitsugaku*) in the Edo period. There is much unique value to be had from the devoted labors of all, even on the, at first, seemingly smallest matters.

Access to source materials

Ideally, each researcher should enjoy full access to all important source materials. At present this is not possible. Many key sources are out of print, and even existing materials are not always easily accessible. Moreover, the majority of the sources are contained in large, expensive reference works, not single, low-cost monographs. This is the case with complete editions of classical treatises in general and for specific sciences. It is also true for the collected works of many important scientists of the past. Generally, the best collections can be found in the libraries of large universities and other research-oriented agencies. For example, the best *wasan* collections are found in the libraries of Tohoku and Kyoto Universities and of the Japan Academy. Astronomical sources are most complete in the Tokyo Astronomical Observatory and the National Diet Library. Good collections of *honzōgaku* materials (botany and zoology) are housed in the Kyōshōku Reference Room of Takeda Laboratories in Osaka and in the Shirai Kōtarō Reference Room of the National Diet Library. The best medical source collections are in the Fujikawa Yū Reference Room of Kyoto University Library and the Fujinami Kōichi Reference Room of Keio University Library.

These are just a few examples. Many other university and institutional libraries have very substantial collections of history of science materials. Certainly one urgent need is the production of a comprehensive bibliography listing all the various holdings of these many institutions. At present, the most useful reference for known works, giving the location of source materials, is the *Kokusho somokuroku* (General catalogue of national documents) published by Iwanami Shoten (1963–72).

2

A History of Universities: an Overview*

From the viewpoint of science history

Nakayama Shigeru

Interrelations between the history of science and the history of universities

What is the significance of the university for science? What has it been? And what ought it to be? These questions are important to the history of science, and this essay endeavors to clarify problems of the history of universities from this vantage point. However, even though I call this an overview, there is yet no established discipline for the study of universities—certainly not among science historians—so what I am presenting is largely a personal view, rather than an objective description of research trends.

Of course, the institution called "university" does not exist solely for the sciences. It is a composite system of many roles and functions which have not always coexisted comfortably. First of all, at the time of its origin in Europe during the Middle Ages it was a kind of guild for the learned professions of theology, law and medicine, as well as an agency for the conferring of academic degrees. Secondly, for those exercising power and authority it became a channel for the recruitment of talent into military and administrative organs, much as in the civil service examination system of traditional China; while for the candidates it served as a gateway to social status and personal advancement. Finally, in the modern educational system the university, as the locus of " higher education," is simply the next stage beyond primary and secondary education. Generally speaking, the first of these roles appeals to medieval historians as an avocational interest; the second is a research theme of

* Published in *Kagakushi kenkyū* (Journal of the history of science, Japan), Vol. 10, No. 100 (1971), pp. 205–207.

sociologists, especially sociologists of education; while the third naturally concerns historians of education.

In contrast to this, the interest of science historians in the university as a locus of scientific research has been limited to a few of the university's functions during a very brief period of time, such as to the faculty of philosophy of the 19th-century German university.

Today, however, people generally are so impressed by the notion of the university as a place for research that the university is sometimes thought falsely to have a monopoly on the production of knowledge. Martha Ornstein's classic work on *The role of scientific societies in the seventeenth century*[1] tried to correct this misconception from the viewpoint of science history. She develops the thesis that during the scientific revolution of the 17th century, informally organized scientific societies were the main arena for the activities of scientists, whereas the established universities played a rather conservative or even reactionary role. Ornstein's argument is quite well known among researchers in the history of science and the attention of subsequent researchers, especially the 'externalists', has been focussed on informal groups of scientists rather than on universities. R.E. Schofield's *The Lunar Society of Birmingham*[2] is one such example.

If the way a scientific paradigm (in Thomas Kuhn's[3] sense of that term) undergoes the process of institutionalization is represented diagrammatically, we have the following:

Paradigm	Supporting groups (proponents)	Reproductive mechanisms (to train scientists)	
			⟶Society
Individuals	Academic societies & research organizations	Universities	

Various paradigms first appear at the level of individual scientists, as candidates for the critical appraisal of others. They then become established at the level of informal groups of researchers or academic societies through a process of critical selection by these groups. Subsequently they find their way into university curricula as means for training the next generation and thus are perpetuated as academic tradition.

The history of scientific thought (the so-called 'internalist approach') deals with the formative stage of the paradigm at the far left of the diagram. The social history of science (that is, the 'externalist approach') tries to establish linkages between the diagram's far left and far right in terms of social and economic factors. It warns against trying to short-circuit connections between the far left and the far right, as Boris Hessen once did. Its recent tendency seems to be concentration on the relationship of the paradigm to the supporting groups, that is, on the formative process of a discipline.

Externalists have always stressed the significance of and the contributions to modern science of the 19th-century German university. But a young sociologist at the University of Heidelberg, Frank R. Pfetsch, has emphasized even more the Gesellschaft Deutscher Naturforscher und Ärzte where scientific research became established even earlier than it did in the university.

Undoubtedly, such an emphasis is justified when one is concerned with the frontiers of a newly-forming paradigm or discipline. However, that emphasis must be reversed when dealing with the transplantation of existing paradigms to colonial areas, as in the case of science in Meiji Japan where first the university system was founded and then groups of scientists were nurtured in it. Here one would have to say that the history of the university is more important in the transplantation phase than in the earlier phase of free inquiry on scientific frontiers.

Although the formulation of paradigms and their supporting groups is an area treated uniquely by scholars of science history and intellectual history, the process of discipline-formation cannot be comprehended without consideration of the process of reproducing scientific offspring. Moreover, at the level of incorporation of scientific disciplines into the university, and because of the admixture of its various functions mentioned earlier, research tools must be extended to those of the sociologist, pedagogist, and others.

A personal view of problems and approaches

Research into the history of science and the history of universities should proceed from the premise that neither established disciplines nor university faculties have any *a priori* reason for existing. Historical research casts all existing things into a relative perspective, a fact all the more clear when comparing the constellation of bodies of learning in different cultural spheres. The Chinese cultural sphere in particular provides evidence against any tendency to absolutize the academic traditions of the West. In China the traditions of rhetoric and logic which undergird Western scholarship were not considered legitimate disciplines. The status of law in the hierarchy of Chinese learning, too, was quite low compared to its status in medieval European universities. Or, in the West astrology was at one time accorded the status of a legitimate discipline, while in Tokugawa Japan mathematics was classified among amusing diversions. One would not, of course, wish to deny differences in the internal character of various bodies of learning, but the standards by which they are evaluated are definitely socially conditioned.

While a particular scientific paradigm first emerges in the mind of a single individual, one cannot assert categorically that conditions within the society to which that person belongs fail to affect the process of its creation. Assuming, even tentatively, that ideas emerge purely by chance, even then paradigms are merely candidates for critical selection or rejection by various supporting groups. This critical work takes place at the level of research groups and academic societies where the more cosmopolitan value criteria of the researchers are dominant. Once a paradigm overcomes this hurdle and gains acceptance in the scientist-reproducing structures of the university and is thence perpetuated as part of scientific tradition, it becomes subject to the special interests of place and time embraced by those structures. More than research organizations, university structures are governed by localized social conditions, so that regional differences operate with a special intensity in the selection and rejection that goes on at this level.

Inquiry into these structures and their mechanisms of selection and rejection may reveal some clues as to why Aristotelian paradigms dominated the intellectual traditions of the West, while in the Chinese cultural sphere the Confucian paradigms of classical study (*ching-hsüeh*) prevailed. Similarly, a comparative consideration of the subsequent fates of various typical scholarly paradigms may lead to classification according to certain social criteria.

Next, one may ask what kind of preparatory work is required for a newly created paradigm to enter the existing reproductive structures of the university, where it becomes institutionalized as a course of study or research program of an institute? In response to this question it seems necessary first to analyze the constituent elements of the preparatory phase required for the organization of a university, into at least two categories: those which are absolutely indispensable and those of only secondary importance. Historically these elements include lectures, examinations, and diplomas as related to employment, all of which are naturally weighted differently according to various historical and social conditions. Why was engineering, for instance, not accepted by the 19th-century German university? Or, why were culinary arts not taught at the university while architecture was? Investigating such questions as these may help determine the necessary and sufficient conditions for the admission of a body of knowledge into the university.

Such inquiries are especially important to assessing the structures of the faculty of philosophy of the 19th-century German university. At that time it contained a number of new disciplines—physics, chemistry, physiology—with new methodologies, disciplines that successively found their place in the university curricula according to methodological definitions. This called for a degree of coexistence, and perhaps compromise, with older disciplines like astronomy and biology, which are defined primarily by their respective objects of study. This particular arrangement, which has lasted to the present day, became fixed at the undergraduate level of the university. Disciplines which failed to secure a foothold at that time have since had great difficulty in gaining a place in the university curriculum.

What changes does learning undergo after it enters the university? Apparently university structures simultaneously promote and limit the development of science. In a word, the university contributes to the normalization and routinization of academic research. A close look at the mechanisms for transmitting knowledge and skill from one generation to another should enable us to single out various elements which define the character of academic work. For instance, the scope of research promoted in universities corresponds roughly to the size of working research units, which in turn are based on the particular human relationships between teachers and students in a particular laboratory. Moreover, because university education is basically lecture-centered, disciplines acquire an introductory, textbook orientation.

The 19th-century German university was not created primarily to further the development of various experimental sciences. At least in its early stage philosophy lectures were its most salient feature, with most groups of students becoming virtually intoxicated by the lofty principles expounded from the lecture podium. This 'theatrical' style of the university was also manifested in the study of nature and in the way knowledge about it was transmitted. The strong impact of this style on other work, especially on science done under the same roof as philosophy, cannot be overlooked. It was not without reason that the rarified and fruitless *Naturphilosophie*, which came to be called "beautiful poetic nonsense," flourished in such an environment. Moreover, the *Privatdozent* system characteristic of the German university, in which a lecturer's income depended upon tuition fees from students attracted to his lectures, necessarily served to confine the study of nature within the framework of a lecture discipline. There was no strong empirical, experimental challenge to the dominance of *Naturphilosophie* until after university-affiliated laboratories emerged in the 1820's and 1830's.

Laboratory science, too, conformed to the same university style. After Justus von Liebig's time university laboratories increased in number and the status of laboratory science rose. However, field sciences which could not be pursued primarily in the laboratory were regarded lightly by university professors as unscientific and amateurish. The status of these disciplines consequently remained quite low. As the university confirmed its position as the authoritative storehouse of knowledge, disciplines unable to enter the university came to be viewed as unorthodox. The term *Wissenschaft* applied only to orthodox fields recognized within the university.

What changes appear in research dispositions after researchers trained in the university form occupational groups? The direction of research shifts in such a way as to conform to the value standards of the professional groups with which the researchers are affiliated, and becomes detached from the evaluative standards of society as a whole. The result, then, is "science for its own sake."

Before science became professionalized, scientists endeavored to make an impression on the learned public in general. Even if they made experiments, they did so only to prove their own theories. But with professionalization they tended to serve only the interests of their own groups and engage only in work endorsed by their scientific colleagues. One could, for instance, obtain an academic degree from one's colleague group by performing an experiment to increase by one unit the degree of precision of the expansion coefficient of copper wire, although such work need have no real significance for one's scientific outlook. If one simply produced more exact data, someone in his profession might be able to make use of it. To have something utilized was to be recognized. So the research data of normal science accumulated very rapidly.

We may also ask what particular sociological significance the specific faculties or departments in Japanese universities have. The faculties of law, theology and medicine in the medieval European universities or the German university's faculty of philosophy were founded to meet the needs of specific professions—lawyers, theologians, physicians and teachers. These one may describe as old-style professions. In contrast to them, disciplines like physics or physiology came into being through modern specialization, in response to the needs of the research professions.

Nowadays the frontiers of research are not necessarily found in the university. However, neither the new research professions nor the older professions can be maintained apart from the university faculties and departments. Likewise, the problematic behavior of the professions cannot be dealt with apart from the university.

Thus far we have concentrated on the normal process by which a paradigm is created and then a profession comes into being to preserve and further it. But can this process be reversed? In fact, it can be. In such a case, a university system is first created, and then in it are formed support groups to service a foreign paradigm transplanted from another culture. The transplantation of Western science to Meiji Japan affords a conspicuous example, and the modernization of most developing countries today seems to follow this same pattern.

Consequently, cases emerge which differ from the 'normal' process. It often happens that in places remote from the institutional inertia of Western universities, ultramodern systems are designed by progressively-minded officials. In Latin American universities, for example, an attempt was made to jump from the structures characteristic of medieval Spain to the most advanced organizational innovations. This was one of the most interesting experiments ever made in the history of universities. But later, as frequently happens, many obstacles were encountered in the process of implementing the innovative designs. Both phases of this process merit careful analysis.

The demerits of establishing an organizational framework ahead of a paradigm in places where no self-generating paradigm exists are often pointed out. For instance, academic societies which are supposed to have emerged simultaneously as supporting groups with a selective and critical stance toward paradigms too often degenerate into clubs of privileged elite wherever the organizational structure is created first. Can we not, though, recognize advantages, even embryonic ones, in places unburdened by fixed paradigms (places without parochial philosophies like Christianity, or entrenched prejudices as found in Newtonian and Cartesian views, in early 19th-century German *Naturphilosophie*, or in dialectical materialism)?

Just one final note. It may be said that scholars fall into two groups: the 'sacred' and the 'secular'. (Or, in more perjorative parlance, 'visionaries' and 'realists.') The history of ideas is generally the province of the visionaries, while the realists tend to concern themselves with the history of structures or systems. The history of universities thus belongs to the realists and demands a sober, adult intellect. Those having this interest are usually persons committed to certain social realities (and as an age group, are graduate students or older). Yet, unlike the history of ideas, this is not a subject which will fulfill educational requirements.

On the other hand, we sometimes become buried in systematic realities to the extent that we lose sight of the kind of external viewpoint from which our notions of that strange and ancient institution called the university, and of its development, could be changed for the better. From time to time it is essential to reopen critical consideration of the meaning of those intellectual enterprises which have characterized the university over a period of several centuries.

References

1. Martha Ornstein, *The role of scientific societies in the seventeenth century* (Chicago, 1938).
2. Robert E. Schofield, "The Lunar Society of Birmingham: a bicentenary appraisal, "*Notes Rec. Roy. Soc. Lond.*, 1966, 21: 144–161.
3. Thomas S. Kuhn, *The structure of scientific revolutions* (Chicago, 1962).

3

The Shifting Center of Scientific Activity in the West*

From the 16th to the 20th century

Yuasa Mitsutomo

From a statistical analysis of scientific achievements recorded in a chronological table for the period 1501–1950 and a survey of outstanding scientists listed in a biographical dictionary, I have discovered some new insights concerning the centers of scientific activity in the Western world. Defining "scientific prosperity" as a period in which the percentage of scientific achievements of a country exceeds 25% of that in the entire world in the same period, I found that the periods of scientific prosperity shifted as follows: Italy (1540–1610), England (1660–1730), France (1770–1830), Germany (1810–1920), and the U.S.A. (1920–present). The average duration of a period of scientific prosperity is about 80 years, with ups and downs which are also discussed below.

In this paper we treat the whole period from the ages of Galileo and Newton to that of Einstein in America, and we also attempt to illustrate graphically the phenomena of scientific activity from the 16th to the 20th century. The nations or national states selected as the units of this historical survey are Italy, England, France, Germany and the U.S.A., with some reference made to the U.S.S.R. and other countries. Our statistical investigation has enabled us to draw some positive conclusions regarding the periods of scientific activity of these nations, periods which are not, of course, as long as the histories of the nations as such.

The first phase of the transition from medieval learning to modern science is the period of the Renaissance when Italian science flourished. In the second phase the center of modern science moved to England, in what may be called the age of

* Published in *Japanese studies in the history of science*, No. 1 (1962), pp. 57–75.

Newton, followed by subsequent shifts to France and Germany. The center of scientific activity is no longer found on the European continent, as from the second quarter of the 20th century it crossed over to the American continent.

I am particularly indebted to Robert K. Merton[1] for his methods of statistical investigation of historical data, and also to J.D. Bernal for his suggestive table[2] from which I gained hints for this paper. Bernal's original Table 8 has only one column for "Centers of Technical and Scientific Activity." In our Table 1, the places listed in Bernal's single column are separated into several columns according to the country of location—Italy, England, France, Germany, U.S.A., U.S.S.R., and others. The basic plan for this paper derives from Table 1.

Materials and methods

Materials.—Five different sources for investigating the phenomena of scientific activity were considered:

1. "Scientific Achievements" recorded in a chronological table, such as Darmstaedter's[3] (in German), and Yuasa's[4] and Heibonsha's[5] (in Japanese).
2. "Representative Scientists" compiled in a biographical dictionary, such as Webster's[6] and Hyamson's.[7]
3. "University Graduates of Science Faculties" (including similar departments).
4. "Scientific Documents" published in periodicals or journals.[8]
5. "Nobel Prize Recipients" (from 1901 to 1960).

A large quantity of objective data must be assembled for the application of scientific methods, and for the elucidation and formulation of laws. Materials in categories 3 and 4 above are not dealt with in this paper, due to difficulties in gathering sufficient objective data for the entire world. Data on Nobel Prize recipients is too limited for comparison with other materials, but is useful for measuring scientific activity of the 20th century.

Method I.—For the present investigation we used Heibonsha's *Kagaku gijutsu shi nenpyō* (Chronology of science and techno-

Table 1 Centers of Technical and Scientific Activity

Year	Bernal's original TABLE 8	Italy	England	France	Germany	U.S.A.	U.S.S.R.	Others
1400	Florence, Venice	Florence			Nurnberg			Bruges
	Bruges, Nurnberg	Venice						
1500	Padua, Cracow	Padua						Cracow
	Copenhagen, Prague							Copenhagen
1600	London, Paris		London	Paris				Prague
1700	Leyden, Paris			Paris				Leyden
	Uppsala, Edingburgh		Edingburgh					Uppsala
1800	Birmingham, Glasgow		Birmingham					
	London, Paris		Glasgow London	Paris				
	Manchester		Manchester					
	London, Paris, Berlin		London	Paris	Berlin			
1900	Cambridge, Paris		Cambridge	Paris				
	Munich, Copenhagen				Munich			Copenhagen
	New England					New England		
	California					California		
	Cambridge, Moscow		Cambridge				Moscow	

Main Current of Scientific Activity

J.D.Bernal, Science in History (London, 1954), pp.930-31.

1833–1836 19 世紀中葉科学・技術史年表

思 想	科 学
1833 独，最初の磁気観測所をゲッティンゲンに設立 (1) このころ天文台，気象台，磁気・地震観測所が各地にでき始める	○ハルミトン（英）虚数の定義としてつの実数の組を提唱 ファラデー（英）電解に関するファラデーの法則(電気素量の考えが胚胎した) (4) ベルツェリウス（典）エチル説を提唱 ペイアン，ペルソー（仏）酵素の発見（変芽中のジアスターゼ) ハワード（英）寒冷前線の概念を導入 L.J.R. アガシ（米）"化石魚類の研究" (~43) を刊行 (5) サン・ナイレール（仏）"動物形態をかえ得る環境の作用程度について" (6) ビューモン（米）消化生理の研究 (7) J. ミュラー（独）"人体生理学提要" (~40) を刊行 (12)
1834 白，ブリュッセル大学創立	○ヤコービ（独）2 変数 4 重周期函数の理論を展開 (8)

Fig. 1. A part of the Heibonsha's *Chronological Table of Science and Technology*, 1956. We used the column of science (科学) classifying items by nations, England (英), Sweden (瑞), France (仏), U.S.A. (米), Germany (独).

Table 2 Number of Items, Heibonsha's *Chronology of Science and Technology* (1956)

Date of scientific achievements	Italy	England	France	Germany	U.S.A.	Others	Total
1501–1550	4	1	1	3	—	4	13
1551–1600	23	7	2	8	—	19	59
1601–1650	15	8	10	15	—	12	60
1651–1700	12	35	15	13	—	18	93
1701–1750	1	17	15	6	2	28	69
1751–1800	7	37	54	32	4	37	171
1801–1850	11	92	144	154	16	41	458
1851–1900	8	106	75	202	33	95	519
1901–1950	10	71	32	128	218	163	622
Total	91	374	348	361	273	417	2064

Table 3 Data on France, 1801–1900, Number of Items Recorded in Heibonsha's *Chronology of Science and Technology* (1956)

Date of scientific achievements	Word (A)	France (B)	France (B/A) %	Moving average % (three points)
1801–1810	90	30	33.3	—
1811–1820	82	32	39.0	36
1821–1830	96	32	33.4	33
1831–1840	109	27	24.8	26
1841–1850	76	14	18.4	23
1851–1860	90	23	25.6	18
1861–1870	100	10	10.0	16
1871–1880	105	13	12.4	12
1881–1890	97	12	12.4	12
1891–1900	105	13	12.4	11
1901–1910	89	6	6.8	—

logy) published in 1956 (Fig. 1). Darmstaedter's *Chronologische Darstellung* is a large and valuable source, but ends with the year 1908.

Heibonsha's *Chronology* contains 2064 entries for the period from 1501 to 1950 (cf. Table 2 and Fig. 1). This overall period is divided into 45 ten-year eras. Chronological data of these eras is classified by nations, and each era's data for each nation is divided by the corresponding world data of the same era to yield a series of percentages. The data of France (1801–1900) and the method of calculation are shown, for example, in Table 3.

Moving averages (three points) of Italy, England, France, Germany and the U.S.A. for the entire period 1501–1950 are drawn in Figure 3. These graphs vividly portray the shifting of the center of scientific activity from the 16th to the 20th century.

Method II.—The second method is entirely different from the first in its materials, though the statistical calculations are similarly performed. We divided two copies of *Webster's biographical dictionary* (1951) into references for 40,000 persons. From one volume we used only the odd-numbered (right-hand)

Per'kin (pûr'kin), Sir Willian Henry. 1838–1907. English chemist; produced mauve, the first synthetic dye (1856); with father and brother, established works near Harrow for manufacture of mauve, thus founding aniline dye industry (1857). Developed process for producing alizarin; first to synthesize coumarin; discovered the Perkin reaction for making aromatic unsaturated acids, such as cinnamic acid (1878); studied relation between chemical constitution and rotation of polarized light in a magnetic field.

英

化 学

Mi'chel·son (mī'kel·s'n), Albert Abraham. 1852–1931. American physicist, b. in Germany; to U.S. (1854); grad. U.S.N.A., Annapolis (1873); studied in Europe (1880–82); professor and head of the department of physics, U. of Chicago (1892–1929). Determined with a high degree of accuracy the speed at which light travels; invented an interferometer for measuring distances by means of the length of light waves; measured a meter in terms of the wave-length of cadmium light for the Paris Bureau International des Poids et Mesures; performed experiment (with E. W. Morley) which showed that the absolute motion of the earth through the ether is not measurable. This demonstration served as the starting point in the development of the theory of relativity. Received 1907 Nobel prize in physics.

米

物理学

Fig. 2. Webster's Biographical Cards, 10×7.5 cm, made of the entries in Webster's *Biographical Dictionary* (1951).

Table 4 Number of Webster's Biographical Cards, Classified According to Nations

Birth year	Italy	England	France	Germany	U.S.A.	Others	Total
1501–1550	5	5	3	5	—	7	25
1551–1600	6	8	5	6	—	9	34
1601–1650	9	18	8	7	—	9	51
1651–1700	4	21	15	5	1	10	56
1701–1750	13	38	41	21	7	27	147
1751–1800	16	84	84	64	26	34	308
1801–1850	23	152	103	167	178	98	721
1851–1900	9	116	47	114	401	123	810
Total	85	442	306	389	613	317	2152

Table 5 Number of Webster's Biographical Cards, Classified According to Departments of Science and Medicine*

Birth year	Mathematics	Astronomy	Physics	Chemistry	Biology	Earth Science	Medicine	Total
1501–1550	7	4	1	—	5	—	8	25
1551–1600	17	7	1	1	4	—	4	34
1601–1650	15	13	8	3	8	—	4	51
1651–1700	18	5	8	1	12	—	12	56
1701–1750	29	22	15	11	42	13	15	147
1751–1800	39	30	49	44	69	47	30	308
1801–1850	76	81	109	110	164	110	71	721
1851–1900	65	72	176	151	178	75	93	810
Total	266	234	367	321	482	245	237	2152

* One scientist is classified in one department; for example, Isaac Newton in physics only.

pages on science and from the other volume only the even-numbered (left-hand) pages on technology, to make up 40,000 individual biographical cards. Samples of these cards (10 × 7.5 cm) are shown in Figure 2. The cards for scientists among these biographical cards were classified by nations, and the cards for each nation (Table 4) were further divided into 40

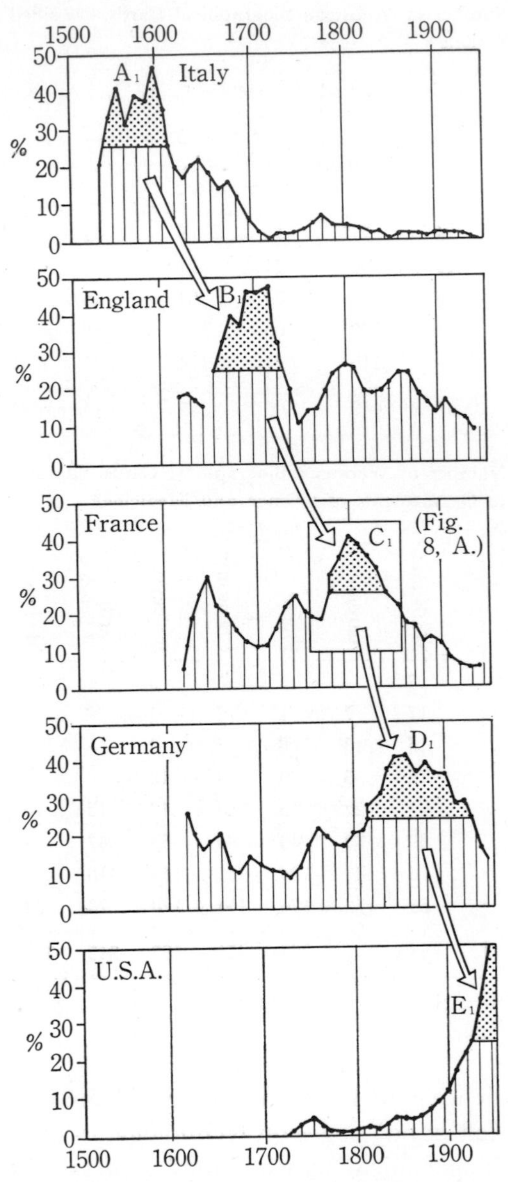

Fig. 3. Method I. The shift of the center of scientific activity from Italy to England, to France, to Germany, and to the U.S.A., obtained by Method I, using Heibonsha's *Chronology of science and technology* (1956).

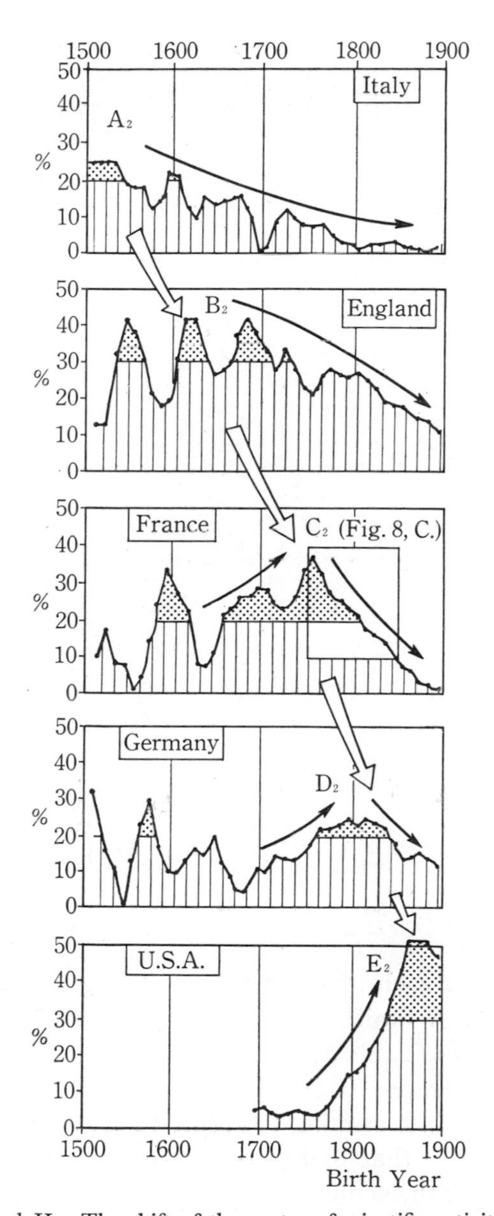

Fig. 4. Method II. The shift of the center of scientific activity obtained by Method II, using Webster's Biographical Cards (cf. Fig. 2).
(Fig. 8, C.)

Table 6 Number of Nobel Prize Winners (1901–1960)

Nation	1901–1910	1911–1920	1921–1930	1931–1940	1941–1950	1951–1960	Total
Group A							
Germany	12	8	8	10	4	4	46
England	5	3	8	7	6	9	38
France	6	5	3	2	—	—	16
Holland	4	1	2	—	—	1	8
Austria	—	1	3	2	1	—	7
Sweden	1	2	2	—	1	1	7
Switzerland	1	1	—	2	3	—	7
Denmark	1	1	3	—	—	—	5
Italy	2	—	—	1	—	1	4
Belgium	—	1	1	1	—	—	3
Hungary	—	—	—	1	—	1	2
Czechoslovakia	—	—	—	—	—	1	1
Finland	—	—	—	—	1	—	1
Spain	1	—	—	—	—	—	1
Group B							
U.S.A.	1	1	3	9	16	27	57
Group C							
U.S.S.R.	2	—	—	—	—	4	6
China	—	—	—	—	—	2	2
India	—	—	1	—	—	—	1
Japan	—	—	—	—	1	—	1
Total	36	24	34	34	36	51	215

groups, one for each ten-year era, according to each scientist's year of birth. A similar classification was made according to specific sciences (Table 5).

Moving averages (three points) of percentages obtained through calculations similar to Method I are drawn in Figure 4. This graph, like that in Figure 3, also makes vivid the successive shifts of the center of scientific activity.

Results

Nobel Prize recipients.—The number of Nobel Prize recipients

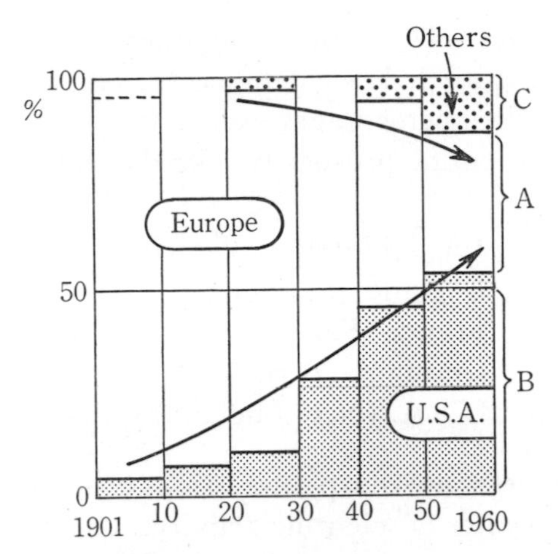

Fig. 5. The percentages of Nobel Prize recipients of A (Europe), B (U.S.A.) and C (Other countries including U.S.S.R.), 1901–1960.

from 1901 to 1960 is 215. These recipients are classified into three groups (Table 6):

A. Europe (excluding the U.S.S.R.)			146
B. U.S.A.			57
C. Others	a. U.S.S.R.		6
	b. Asia	India	1
		Japan	1
		China	2
	c. South America		1
	d. Australia		1

European recipients make up 68% of the world total, and those of the U.S.A. 26%. These percentages are the mean values for the sixty-year period 1901–1960. This entire period is divided into six ten-year eras, and the number of recipients of the three groups (A, B, C) for each era are divided by the world total for each corresponding era. The results are shown in Figure 5. This graph clearly indicates that the center of scientific activity, especially high-level creative activity, is now in the U.S.A. This center remained in Europe from 1901 to 1930, as the mean value for Europe during that time span was about 90%.

Chronological table.—The results obtained by Method I are illustrated in Figure 3. If we define a period of scientific prosperity as one when percentages exceeded 25% of world totals, the periods of scientific prosperity shifted as follows:

1540–1610	Italy (Florence, Venice, Padua)
1660–1730	England (London)
1770–1830	France (Paris)
1810–1920	Germany (Berlin)
1920–present	U.S.A. (New England, California)

Figure 3 provides some sharp impressions concerning the transference of the center of scientific activity. The main points are as follows:

Italy: The first country to achieve scientific prosperity was Italy, where the main stage of the Renaissance first unfolded and great scientists like Leonardo da Vinci and Galileo Galilei come into prominence.

England: Scientific prosperity in England ended about 1730, when Isaac Newton (1642–1717) died. The rapid decay of scientific activity after Newton's death provides a striking contrast to Bernal's table (cf. Table 1). The data treated statistically in Figure 3 includes only scientific achievements, not technical achievements. If the technical data in Heibonsha's *Chronology* (the right-hand pages) is added to the scientific data, the period of combined technical and scientific prosperity continued to the end of the 19th century (Fig. 6). The number of Nobel Prize recipients (Table 6) indicates the creativity of British scientists in the 20th century, but it is not unwarranted to say that the period of scientific prosperity in England had ended. This conclusion can be drawn from Figures 3, 4 and 5.

France: The decline of French science is clear from the era columns in the table of Nobel Prize recipients (Table 6). French recipients dropped from six in the first decade to only two by the fourth, with no recipients in the last two decades in the period covered. French science flourished in the ages of the encyclopedists, the French Revolution, and Napoleon I; it

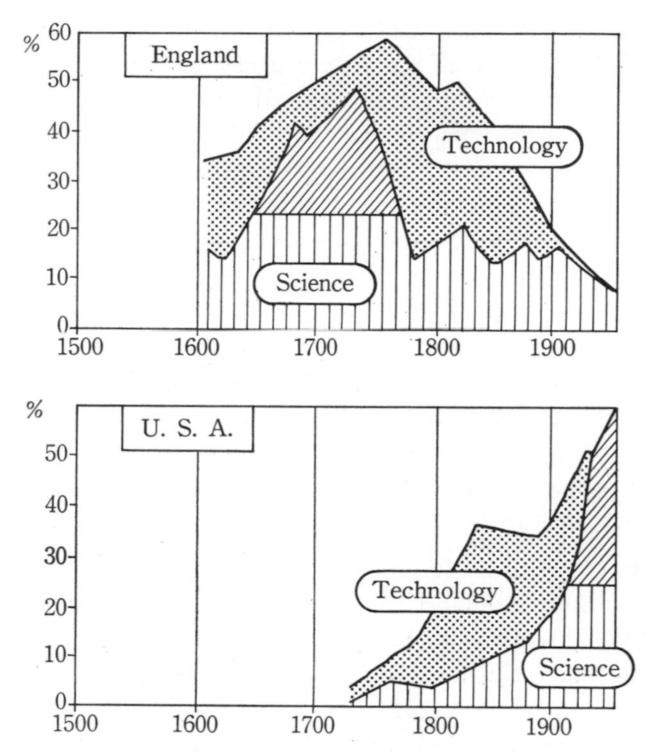

Fig. 6. The "Science" sectors are the same as in Fig. 3. The "Technology" sectors were obtained by a similar method, using the even-numbered (left-hand) pages of Heibonsha's *Chronology of science and technology* (1956). The order of science and technology is inverted in the case of England, where developments in technology came after the scientific revolution, and in the U.S.A., where this sequence was reversed.

dcclined almost in a straight line with the lapse of years after Napoleon I to De Gaulle, a period of about 150 years.

Germany: The period of scientific prosperity of German science continued from the year of the founding of Berlin University in 1809 to the time of the formation of Hitler's Nazi Party in 1921. Germany at present is separated into two parts, West and East, neither of which is likely to become the main stage of modern science. But the center of scientific activity in the 19th century undoubtedly was in Germany, as can be seen in Figures 3 and 4.

U.S.A.: The scientific prosperity of the U.S.A. began in 1920. Until the end of the 19th century American science was rather poor in comparison with its European counterpart. American science experienced only very slow progress during the period 1501–1900, some four hundred years of great difficulty and dependence on European science. In our own century the center of scientific activity has clearly shifted from Europe to the U.S.A., as is shown in Figures 3, 4 and 5.

Biographical cards.—The results obtained by Method II are illustrated in Figure 4. The scale of *abscissa* of this graph is the birth year. If the "acme" of a scientist's productivity is estimated at age forty, then the periods of scientific prosperity will run forty years later. Figure 3, then, would have the following periods (names in parentheses are those of outstanding scientists born in the respective periods):

1500–1570	Italy (Galileo)
1620–1690	England (Newton)
1730–1790	France (Lavoisier, Laplace)
1770–1880	Germany (Gauss, Liebig, Helmholtz)
1880–	U.S.A. (Hubble, Lawrence)

These periods are designated A_1, B_1, C_1, D_1, E_1 in Figure 3. If we compare this set A_1—E_1 with its corresponding set A_2—E_2 in Figure 4, we find that these two graphs represent the same phenomena concerning scientific activity.

Discussion

We can classify the five instances of transference of scientific activity during the period from the 16th to the 20th century into the following three types (cf. Figs. 3 and 11):

Italy: the origins of modern science.

England, France and Germany: The rise and fall of scientific activity over a period of about 300 years.

U.S.A.: Science at present, and the question of the future.

Problems related to these three types include: the origins of modern science, treated, for example, by Herbert Butterfield;[9]

and the introduction of the concept of "scientific revolution," a problem still awaiting solution.

In discussing the rise and fall of scientific activity, we have chosen French science as a typical case among the three types. To discuss the future of American science is most difficult, though we touch on this question below.

Analysis of French scientific prosperity.—The laws which regulate this phenomena lie not within the scope of scientific knowledge as such, but in the social structure of France at that time. There are two possible approaches to this phenomena, historical and sociological; our discussion is made from the viewpoint of the "sociology of science," which is a new facet of sociology. What happened in French history in the first half of the period of French scientific prosperity, that is, from 1750 to 1800? During this period the social structure of France was in a state of "anomy,"[10] as the following events indicate:

1751–1772: The publication *Encyclopédie, ou dictionnaire raisonné des sciences, des arts et des métiers*, compiled by Diderot and d'Alembert, was placed on the Index of prohibited books.

1755: Rousseau's *Discours sur l'origine de l'inégalité parmi les hommes* (1754) was published.

1756–1763: The Seven Years' War, in which France played a leading role, was waged.

1778–1783: France intervened in the American War of Independence on the side of the colonists.

1789–1795: The French Revolution occurred.

1799–1804: The Consulate, a new system, was established in conjunction with Napoleon's rule.

The first half of the period of French scientific prosperity was a time of anomy in which French science advanced rapidly. If we define the first half of a period of scientific prosperity as the time of scientific revolution,[11] we find that such times in England and Germany were also times of anomy, as shown in Figure 7. There seem to be necessary relations between political revolution and scientific revolution. The main incidents of scientific revolution in England, the founding of the Royal Society (1660) and the publication of Newton's *Principia*

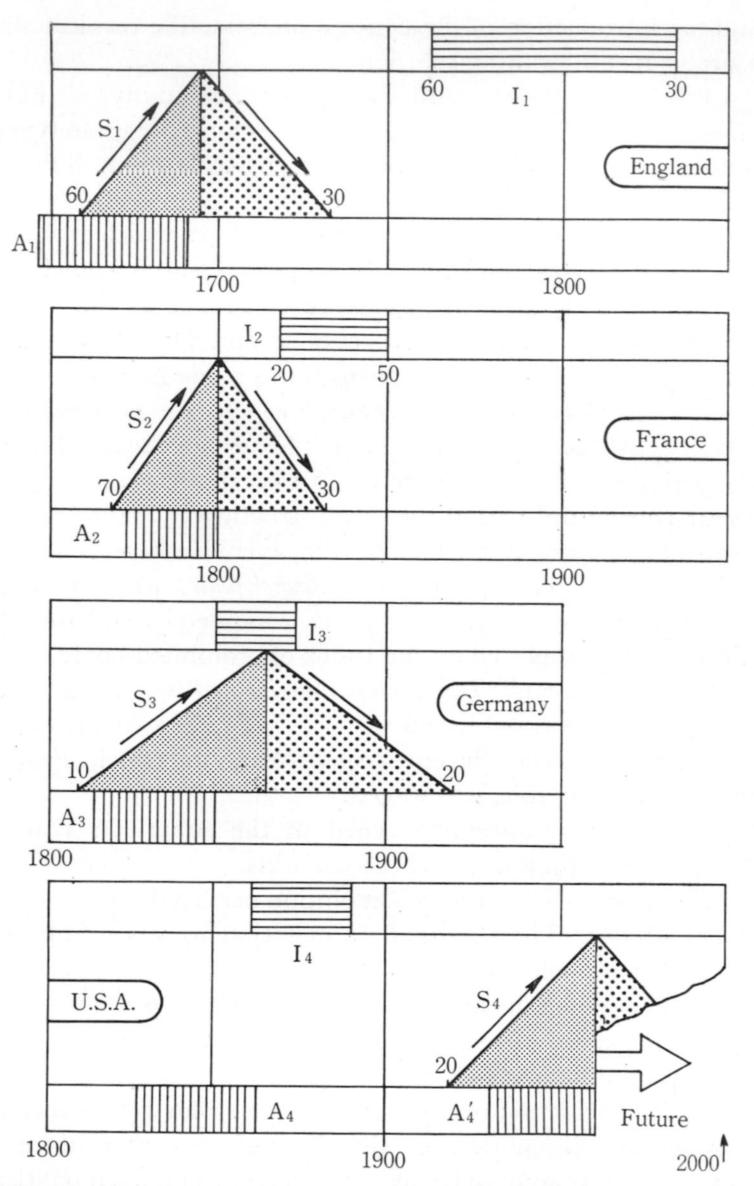

Fig. 7. The relations among Scientific revolution (S), Anomy (A), and Industrial revolution (I) are expressed in these sketches. Representative incidents of A_4 in the U.S.A. are the stock market crash (1929), World War II (1938–1945), and the Korean War (1950–1953). More recent Anomy results from the international showdown between the U.S.A. and the U.S.S.R.

(1687), happened in the time of anomy from the Puritan Revolution (1649) to the Glorious Revolution (1688). In Germany, the Revolution of 1848–1849 broke out in the first half of the period of German scientific prosperity, that is, its period of scientific revolution from 1810 to 1865.

What caused the fall of the flourishing French science? What happened between 1800 and 1830, in the second half of the period of French scientific prosperity? Perhaps there were many reasons for the declining course of scientific activity in France. In this paper we would only point out one of the most important causes, one found in the statistical procedures used for the biographical cards (Fig. 2). It is a phenomenon which may be called the "aging" of the scientific community. The aging of the group of active scientists may be one of the effective causes that brought about the fall of French science in the 19th century.

The phenomenon of aging in France is indicated in Figures 8 and 9. Graphs A and C of Figure 8 are the same as the parts for France for the period 1730–1900 in Figures 3 and 4 respectively. Graph B of Figure 8 was obtained by the following procedure: (a) The French scientists in *Webster's biographical dictionary* who lived or died between 1750 and 1900 are chosen. (b) The number of scientists by age group for every ten years, 0–9, 10–19, 20–29, . . . , is counted in the eight historical sections, 1750, 1770, 1800, 1850, . . . , which are indicated by the signs α, β, γ, δ . . . . The results are shown in Table 7 and Figure 9. (c) In every historical section, α, β, γ, δ . . ., the average age of all scientists in existence listed in *Webster's biographical dictionary* are calculated (Table 7).

In spite of the increase in the absolute number of scientists, if the percentage of French scientists in the world decreases (Fig. 8, graph C), then scientific activity will decrease gradually by the aging of the group of scientists. The fall of scientific activity in England after Newton's death (1727), and in Germany after the organization of the Nazi Party (1921), may be attributed to the respective aging of the groups of scientists.

The future of American science.—Historians rarely discuss the

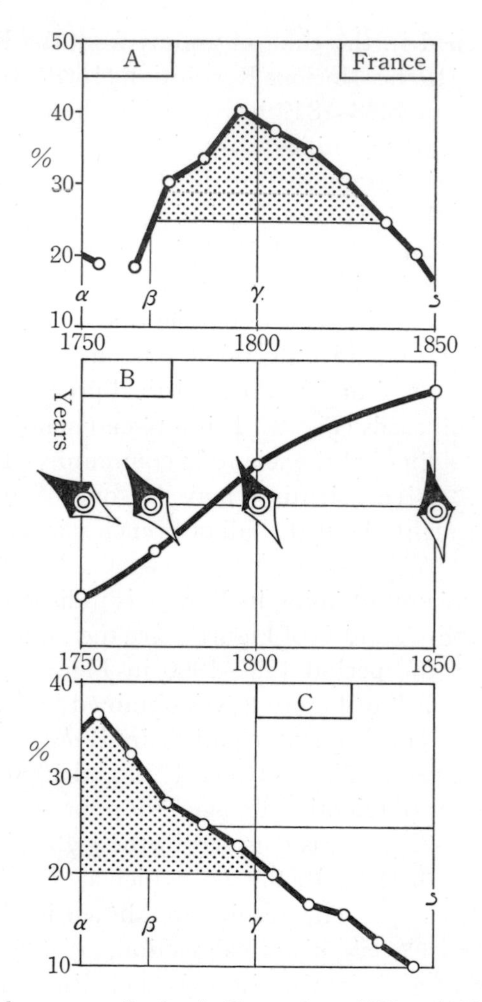

Fig. 8. The phenomena of aging in France from 1730 to 1900. Graphs A and C are the respective parts C_1 and C_2 in small frames in Figs. 3 and 4. The Graph B is obtained from Table 7.

future of society, although natural scientists (meteorologists, astronomers, physicists, chemists, etc.) regularly forecast natural phenomena according to natural laws. We have developed two parameters from which to project the future of American science: the longevity of periods of scientific prosperity in Italy, England, France and Germany; and the rates

Table 7 Data on France, the Aging of Groups of Scientists

Birth year	1890–99	1880–89	1870–79	1860–69	1850–59	1840–49	1830–39	1820–29	1810–19	1800–09	1790–99	1780–89	1770–79	1760–69	1750–59	1740–49	1730–39	1720–29	1710–19	1700–09	1690–99	1680–89	1670–79	1660–69	Total
1750–59																13	10	3	7	4	2	3	.	1	43
60–69															16	13	10	3	7	4	1	1	.	.	55
70–79														11	16	13	10	3	7	2	1	.	.	.	63
80–89													19	11	16	13	10	3	5	1	.	.	.	.	78
90–99												16	19	11	15	13	9	3	3	.	.	.	.	.	89
1800–09											22	16	19	11	15	11	6	2	1	.	.	.	.	.	103
10–19										19	22	16	19	9	10	10	4	.	.	.	.	.	.	.	109
20–29									22	19	22	16	18	9	10	8	.	.	.	.	.	.	.	.	124
30–39								20	22	19	21	15	17	8	7	2	.	.	.	.	.	.	.	.	131
40–49							24	20	21	19	20	14	13	2	.	.	.	.	.	.	.	.	.	.	133
50–59						18	24	20	21	19	17	11	10	.	.	.	.	.	.	.	.	.	.	.	140
60–69					19	18	24	20	20	16	14	5	3	.	.	.	.	.	.	.	.	.	.	.	139
70–79				16	19	18	24	20	19	12	10	2	.	.	.	.	.	.	.	.	.	.	.	.	140
80–89			11	16	19	18	24	19	14	8	3	1	.	.	.	.	.	.	.	.	.	.	.	.	133
90–99		1	11	16	19	17	20	17	10	.	.	.	.	.	.	.	.	.	.	.	.	.	.	.	111
1900–09	1	1	11	16	19	16	18	11	1	.	.	.	.	.	.	.	.	.	.	.	.	.	.	.	94

of increase in the number of Nobel Prize recipients in the U.S.A. and the U.S.S.R.

The longevity of periods of scientific prosperity in Europe were as follows:

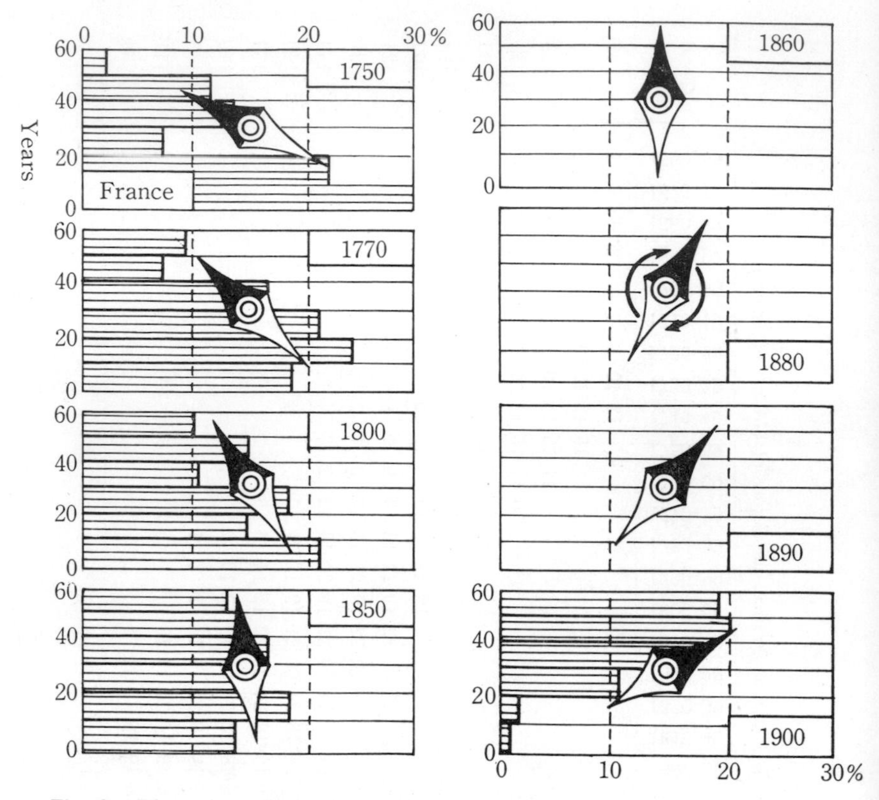

Fig. 9. The aging of scientists in France.

Italy	70 years (1540–1610)	Mean value = 80 years
England	70 years (1660–1730)	
France	60 years (1770–1830)	
Germany	110 years (1810–1920)	

If this pattern holds true also for the U.S.A., then its scientific prosperity, which began in 1920, will end in the year 2000, as indicated in Figure 11. This projection is, of course, dependent upon the "working hypothesis" that American science will be regulated by the same laws that operated in Europe., including the scientists' aging phenomenon (Fig. 10; Tables 8 and 9).

Figure 12 is sketched by making an extrapolation of Figure 5 showing the ratio of Nobel Prize recipients. We cannot dis-

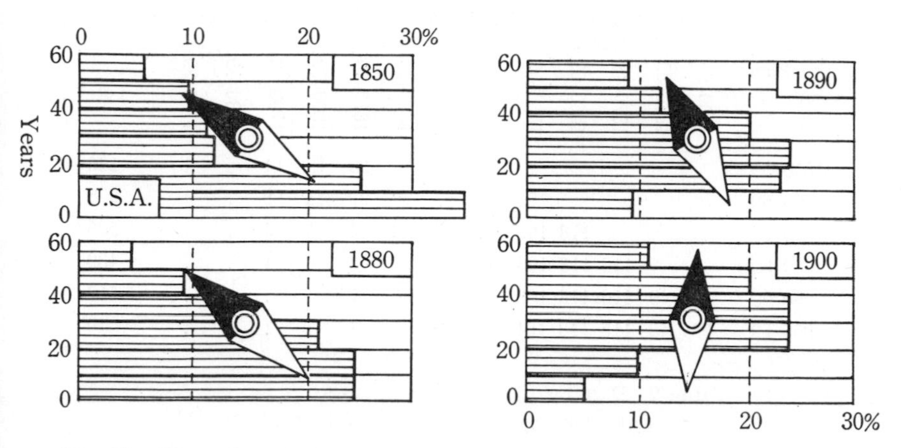

Fig. 10. The aging of scientists in the U.S.A.

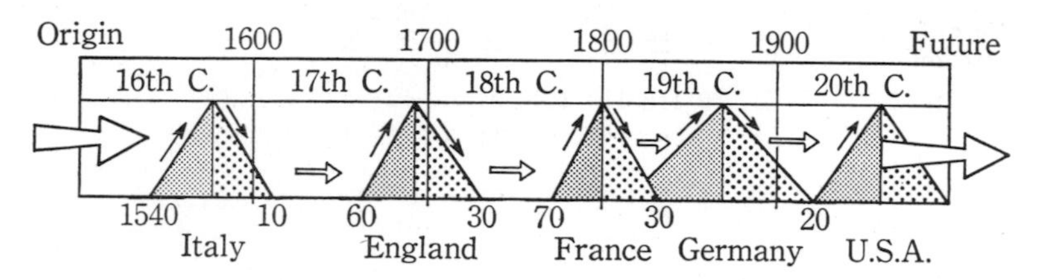

Fig. 11. The shifting sequence of scientific prosperity: Italy (1540–1610), England (1660–1730), France (1770–1830), Germany (1810–1920), and the U.S.A. (1920-present).

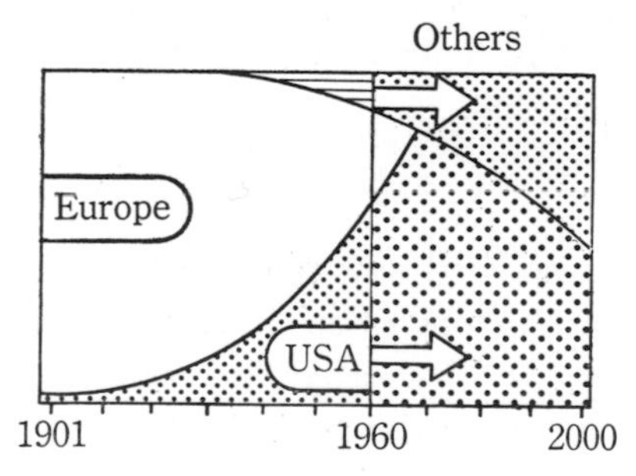

Fig. 12. An extrapolation of the percentages of Nobel Prize recipients in Fig. 5.

cuss the future of the U.S.A. without considering that of the U.S.S.R. The future of other countries (Fig. 12, including the U.S.S.R.) will, in turn, be determined by the future of the

Table 8 Data on U.S.A., the Aging of Groups of Scientists

Birth year	1890–99	1880–89	1870–79	1860–69	1850–59		1830–39	1820–29	1810–19	1800–09	1790–99	1780–89	1770–79	Total
1850–59						64	47	22	20	16	11	5	1	186
60–69					103	64	47	22	20	13	8	.	1	278
70–79				118	103	64	47	22	16	10	5	.	.	385
80–89			116	118	103	64	45	21	13	8	1	.	.	489
90–99		49	116	118	103	63	43	18	7	.	.	.	.	517
1900–09	23	49	116	118	101	57	40	5	1	.	.	.	.	510

Table 9 Data on the Aging of Scientists in the U.S.A. (The Graphs for Fig. 10 are Obtained from C and D.)

Year of section \ Birth year	1890–99	1880–89	1870–79	1860–69	1850–59	1840–49	1830–39	1820–29	1810–19	Total (B)
1900	23	49	116	118	101	57	40	5	1......(A)	510
	5	10	23	23	20	11	8	1 ...C%=(A/B)		
Years	0–10	11–20	21–30	31–40	41–50	51–60	.........D			

U.S.A. Of course, behind this there are two "working assumptions," that the number of Nobel Prize recipients is a valid indication of scientific activity; and that the curves for the U.S.A. and other countries can be extrapolated as far into the future as the year 2000.

Conclusion

Herein we have described new phenomena to be called Yuasa's Phenomena in the history of science, concerning the center of scientific activity, as illustrated in various graphs based on statistical procedures for the period from the 16th to the 20th century. Since one of the main causes of the decline of scientific activity appears to be the aging of the scientific community, perhaps the duration of scientific activity in a given nation can be prolonged by well-planned scientific education and efficient administration.

References

1. Robert K. Merton, "Science, technology and society in seventeenth century England," in *Osiris*, Vol. 4, Part 2 (1938), pp. 360–632.
2. J. D. Bernal, *Science in history* (London, 1954), pp. 930–931, Table 8.
3. Ludwig Darmstaedter, *Handbuch zur Geschichte der Naturwissenschaften und der Technik* (Berlin, 1908, reprinted 1961), pp. 1–1070.
4. Yuasa Mitsutomo, *Kagaku bunka shi nenpyō* (Chronology of science and culture; Tokyo, 1950, new edition 1960).
5. Heibonsha's *Kagaku gijutsu shi nenpyō* (Chronology of science and technology; Tokyo, 1956).
6. *Webster's biographical dictionary*, A dictionary of names of noteworthy persons with pronunciations and concise biographies (Springfield, U.S.A., 1951).
7. Albert M. Hyamson, *A dictionary of universal biography of all ages and all peoples* (London, 1916, 2nd edition, 1951).
8. Derek J. Price, "Quantitative measures of the development of science," in *Archives internationales d'histoire des sciences*, Vol. 14 (1951), pp. 85–93.
9. Herbert Butterfield, *The origins of modern science 1300–1800* (London, 1949).
10. Robert K. Merton, *Social theory and social structure* (Glencoe, U.S.A., 1949), Chaps. IV and V.
11. The History of Science Society of Japan, ed., *Kagaku kakumei* (Scientific revolution; Tokyo, 1961).

4

Statistical Approaches to the History of Science*

Yagi Eri

My interest in the history of science began with the "Bikini shock" caused by an American H-bomb experiment in 1954, when Japanese fishermen were injured by radioactive fall-out. I came to question my future as a physicist who might produce something harmful to ordinary citizens. Therefore, I started searching for a style of learning which could direct the development of science toward human welfare. Thus I decided on study of the institutional history of science, or more specifically, the external approach to the history of physics.[1]

At that time in Japan a book by Amano Kiyoshi on the origin of the theory of heat radiation[2] was highly regarded by historians of physics. Amano showed in detail the close relation between the theory of heat radiation and the quantum theory, on the one hand, and the rise of German industry on the other. While I admired Amano's work, it forced me to ask myself whether it is sufficient to treat science history only in a chronological fashion. Each historical fact, of course, occurs in a time sequence, but why should the study or writing of history stick to chronological order? My search for a method of studying the history of science focussed on that question.

Because of my family background (my father was a pioneer of biometry in Japan) and my educational background, a mathematical approach seemed most suited to me. I was attracted to Derek Price's paper on the quantitative measure of the development of science,[3] and to T.J. Rainoff's paper on the wave-like fluctuations of creative productivity in the development of physics in Western Europe in the 18th and 19th centuries.[4] In 1959 I began to apply Price's method to an analysis of the growth of physics in Japan. At that time I

* Written June 1973

considered Price's method useful for a general analysis of science (especially physics) from a macro-viewpoint, and Rainoff's method, since it dealt more with the internal development of science, as more effective from a micro-viewpoint.

The first of my papers presented here on "The statistical analysis of the growth of physics in Japan" was published in part as an *Abstract paper presented at the New York meeting of the A.A.A.S.*[5] in 1960, and in part in the *Proceedings of the Xth International Congress of the History of Science, 1962.*[6] Here I discussed the structure of normal exponential growth as well as rate of growth of science in Japan. I hoped that the results of my study might provide a prototype for the indigenous training of scientists in developing countries of Asia, Africa and other parts of the world. I recognized, of course, the importance of political independence and adequate financing in developing countries. My intention was to portray the Japanese case as the simplest possible example for such countries, although they had not necessarily expressed any desire to follow the Japanese pattern, and indeed some even rejected it as a model. However, it is still a fact that the Japanese case is one way through which developing countries could reach international levels in science.

The second paper on "An application of a type of matrix to analyze citations of scientific papers" was published in *American documentation.*[7] It analyzes the relations between papers by members of a Japanese group of nuclear physicists (the Elementary Particle Theory Group) in the period 1935–1950 by use of a type of matrix. The use of a type of matrix for the analysis of a series of scientific papers was suggested by Derek Price,[8] while the techniques of the matrix analysis had been developed for the study of group structure in the fields of sociology and psychology. A suitable method was developed in my paper for the purpose of analyzing the Japanese group of nuclear physicists. (Kaneseki Yoshinori investigated the history and background of the group.[9])

My main interest in the group was as follows: it had already manifested itself as a group with strong internal relations in some historical documents (partly because of World War II) but how about the relations of its members to foreign works? The result of my analysis indicated that the papers of the group's members are closely connected within themselves, but not completely confined to that scope.* It is hoped that other groups of scientific papers will be analyzed by this method.

The third paper "On the measurement of science,"[10] published as a general study of statistical approaches, was based on my report to a seminar conducted by Stevan Dedijer at Yale in 1963. It gives a classification of possible types of measurements of science and a brief analysis of governmental documents of the U.S.A., Japan and some developing countries, documents using various measurements of science. This study shows that measurements concerning scientists are most frequently used for national research policy decisions, and that the percentage of gross national expenditure for research and development in relation to national income is the most common measurement.

The fourth paper on "Science education in tropical African universities after World War II"[11] was published as a cooperative work with (Mrs.) Fujita Chie in 1968, in line with my interest in methods for developing science in ways applicable to developing countries from the standpoint of their own welfare rather than from that of developed countries, including Japan. The investigation of the conditions of science education in tropical Africa seemed important, since few works on that subject had been published to date, except for some Unesco publications. As we could not discuss the exponential growth of science, we made some elementary discussions with the aid of simple statistical data. In order to show the nature of African science and technology, we focussed on the structure of science and engineering departments at each university. This meant that an external approach, including statistical methods, was not sufficient in itself; we were forced

* The figures which show the data of actual connections have been omitted from this paper in this book.

to dig more deeply into science as such.

I am not completely satisfied with these statistical studies, as they are more or less surface approaches to the development of science, or even just physics. The problem has always been how to make a further generalization if one assumes that the first approach is a statistical one. My interest in the history of science continues to the present, and I sometimes dream of building a bridge between the external and internal approaches.

Notes

1. For my work on this, see my joint paper with Itakura Kiyonobu on the establishment of the Institute of Physical and Chemical Research, pp. 158–201 of this book.
2. Amano Kiyoshi, *Netsufukusharon to ryōshiron no kigen* (The origin of the theory of heat radiation and quantum theory; Tokyo, 1943).
3. Derek Pricc, *Archives internationales d'histoire des sciences*, No. 14 (1951), pp. 85–93.
4. T. J. Rainoff, *Isis*, Vol. 12 (1929), pp. 287–319.
5. Yagi Eri, "The growth of modern science in Japan," *Abstract paper presented at the New York meeting of the A.A.A.S., 1960.*
6. Yagi Eri, "Comments on a paper by Prof. Watanabe," *Proceedings of the Xth International Congress of the History of Science, 1962*, Vol. 1 (Paris, 1964), pp. 208–210.
7. Yagi Eri, "An application of a type of matrix to analyze citations of scientific papers," *Journal of the American Society for Information Science* (formerly *American documentation*), Vol. 16, No. 1 (1965), pp. 10–19.
8. Suggested by Derek Price at a seminar on the Sociology and Politics of Recent Science, at Yale in 1963.
9. Kaneseki Yoshinori, "The Elementary Particle Theory Group," in this book, pp. 221–252.
10. Yagi Eri, "On the measurement of science," *Scientific papers of the College of General Education, Tokyo University*, Vol. 14, No. 1 (1964), pp. 129–138.
11. Fujita Chie and Yagi Eri, "Science education in tropical African universities after World War II," *Japanese studies in the history of science*, No. 7 (1968), pp. 143–157.

(1) The Statistical Analysis of the Growth of Physics in Japan*

Yagi Eri

In order to measure the growth of modern science, including physics, in Japan, statistical studies have been made using as indices the total number of doctors of science at Japanese universities and the membership of the Physico-Mathematical Society of Japan.

The first graph (Fig. 1) is the total number of doctors of science at Japanese universities from 1888 to 1956, plotted on a logarithmic scale. In the period from 1898 to 1956, the line is almost straight. This means that the growth was exponential during this period. One may say that the number of the doctors, starting with 18 in 1888, doubled every ten years, reaching 2,410 by 1956. An equation for the growth during the period may be written as follows: 1898–1956: number of doctors= 18 *exp.* (0.07 n). In this equation, n is the number of years since 1888.

The second graph (Fig. 2) is the membership of the Physico-Mathematical Society of Japan from 1877 to 1945. This growth, from 1888 to 1945, is also exponential. One may say that the membership of the society, starting with 37 in 1877, doubled every 10.8 years, except for the period from 1912 to 1918, when there was a drop in the rate of growth, mainly because of World War I. The equations may be written as follows:

1888–1911: membership=37 *exp.* (0.064 n);

1919–1945: membership=33 *exp.* (0.064 n). Here n is the number of years since 1877.

Derek Price has maintained that from measurements of the development of science in West using national statistics, scientific publications, and figures provided by learned societies, such as the number of the *Physics abstracts* published in each

* See references 5 and 6, p. 107.

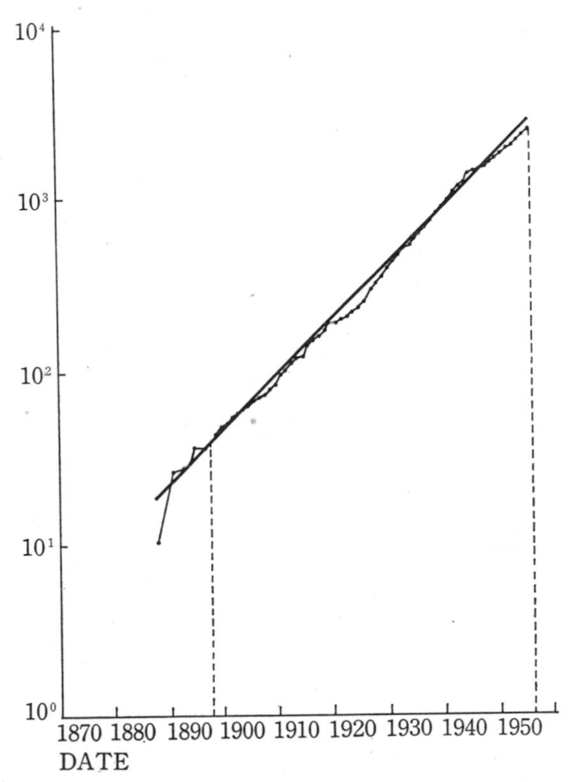

Fig. 1. Total number of doctors of science at Japanese universities, 1898–1957.
1898–1956: Number of doctors $= 18 \ exp \ (0.07 \ n)$
n is the number of years elapsed from 1888. (logarithmic scale)

year, the following three important conclusions can be drawn:[1]
(1) Nearly all curves of growth show the same trends.
(2) The growth is exponential.
(3) The constant of the exponential curve is such as to effect a doubling in size in an interval of the order of 10–15 years.
These findings are now confirmed for the growth of modern science, including physics, in Japan.

In order to show how the constant exponential growth of physics was caused in Japan, the third graph (Fig. 3) was made. This graph indicates the process of domestic production of physicists in Japan.

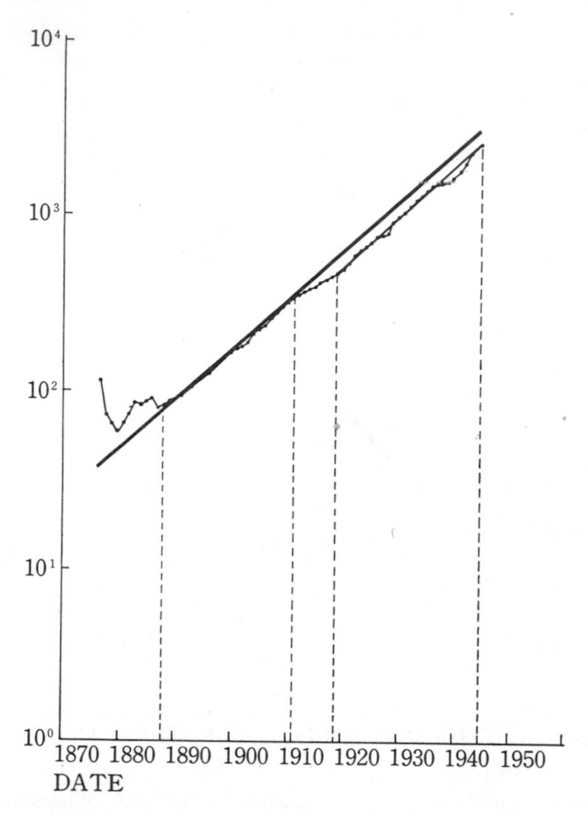

Fig. 2. Membership of the Physico-Mathematical Society of Japan, 1877–1945.
1888–1911: Membership=37 *exp* (0.064 *n*)
1919–1945: Membership=33 *exp* (0.064 *n*)
n is the number of years elapsed from 1877. (logarithmic scale)

The foreign teachers and the first batch of foreign-trained Japanese physicists are together designated Group I, numbering 22 physicists. They began to arrive in 1869 and reached a maximum number of 15 in 1878. By World War I they had left Japan, died, or become inactive.

The students of these innovators may be referred to as Group II. From 1877 they gradually rose to a stable level of about 20 physicists, eventually declining by mortality, etc., during the mid-1920s.

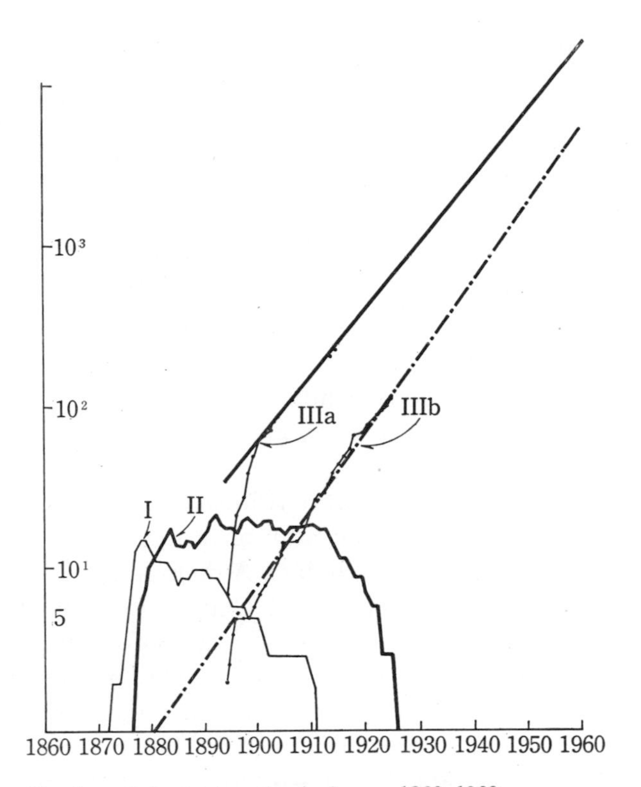

Fig. 3. Number of physicists active in Japan, 1860–1960.
I. Foreign professors and first batch of foreign-trained Japanese physicists.
II. Japanese students of Group I, first generation.
III. Japanese students of Japanese physics teachers (Group II); second and subsequent generations:
a. graduates of university physics courses.
b. students continuing for D.Sc.
(logarithmic scale)

Following the emergence of Group II, Japanese students could for the first time be trained at home and in their own language by Japanese teachers. This crop of home-grown physicists may be designated Group III. Whether we take the crude number of graduates of the university physics courses (IIIa), or the more distinguished group who continued their studies for a D.Sc. (IIIb) which required over 10 years of research work, the result is the same—exponential growth with a doubling period of about seven years and extending almost without change from 1900 to the present. Thus in Japan there are now about 1,000 physicists holding this degree, and the total population of physicists (membership of the Physical Society of Japan) is several times greater in number.

It is interesting that the present statistical data shows the historical difference between the early period of the first few, and the later period of constant exponential growth. The early period, lasting from about 1870 to 1895, is that of "colonial science," dominated by the attitudes of foreign teachers and directed towards such local studies as earthquakes and geomagnetism, local natural history, and the properties of Japanese magic mirrors.

Most Japanese physicists who belonged to Group I and Group II worked as scientific administrators and educators rather than as international scholars. A dictionary of Japanese equivalents of English, French and German technical terms in physics was also published by them in 1888.[2] Its publication greatly improved scientific communication and education in Japan by providing a common language. Not until Western science had been maintained in Japan for nearly a quarter of a century and had started on its serious exponential expansion did the pattern change.

The present Japanese tradition of modern experimental and theoretical physics was not formed until the 1920's and 1930's. The appearance of Japan as a genuinely separate entity in the international world of science seems to date also from this period. Therefore, the true emergence of modern physics in Japan stems from these home-grown physicists.

Japan's case may be taken as a prototype for the underdeveloped countries in which Western physics is now being introduced.

References

1. Derek Price, "Quantitative measure of the development of science," *Archives internationales d'histoire des sciences*, Vol. 4 (1951), p. 85.

———, "The exponential curve of science," *Discovery*, Vol. 15 (1954), p. 240.
———, *Little science, big science* (New York, 1963).
2. Society for Translation of Physics (Butsurigaku Yakugo Kai), *Dictionary of Japanese, English, French and German physical terms* (1888).

(2) An Application of a Type of Matrix to Analyze Citations of Scientific Papers*

Yagi Eri

Introduction

A scientific paper has a series of citations belonging to the previous literature. In order to determine a linkage pattern of scientific papers through their citations, an analysis of a group, whose papers are frequently cited by themselves, is useful. It is possible to investigate the internal connections among papers in such a group by examining their citations. The use of a type of matrix for the analysis was suggested by Derek Price,[1] while the techniques of the matrix analysis had been developed for the study of group structure in the fields of sociology and psychology.[2]

The main purpose of this study is to investigate the measurements by which the internal connections of a group of scientific papers are measured.[3] The techniques of using the type of matrix, suggested by Derek Price, will be further developed to carry out the analysis of a certain type of group of scientific papers. The collected papers by members of the Japanese Group of Nuclear Physics will be used as an example.** It has been suggested that these collected papers must have close internal connections among themselves because of the character of the Japanese Group of Nuclear Physics.

In Part I of this paper the type of matrix to analyze citations of scientific papers will be shown in general.

In Part II of this paper several new types of measurements

---

* Appeared in the January 1965 issue of *American documentation* (presently *Journal of the American Society for Information Science*). Research for this paper was supported by United States Public Health Service Training Grant No. 2G–450, at the Department of History of Science and Medicine, Yale University, 1960–1963.

** The Japanese Group of Nuclear Physics was founded by Nishina Yoshio in the 1930's and has been developed by Yukawa Hideki, Tomonaga Shin'ichirō, Taketani Mituo, Sakata Shōichi, and other Japanese physicists. (See III–4, where this group is called the Elementary Particle Theory Group. -Ed.)

will be proposed to indicate the internal connections of a group of scientific papers by applying the above type of matrix to the collected papers of the Japanese Group of Nuclear Physics.

Part I. A type of matrix to analyze citations of scientific papers

The construction of a type of matrix to analyze citations of scientific papers in a group will be shown in this section. The matrix is constructed by listing the papers in chronological order both along the rows and down the columns. The rows represent the papers cited and the columns the source papers.

Let Xij be called an element of the matrix, Xij being the element in the i row and j column. The row suffix i ranges over the values, 1,2 . . ., n, and the column suffix j over the same values, 1,2 . . ., n. Here n is the total number of papers in the group. The matrix as a whole will be denoted by X.

In this matrix X, one (1) means a citation and zero (0) no citation, so that if the element $X_{12}=1$, this means that paper No. 2 cites paper No. 1; if $X_{12}=0$, paper No. 2 does not cite paper No. 1. This is called a matrix of one-link connections, which indicates direct links among n scientific papers. Figure 1 is the matrix of one-link connections. In this matrix X, the

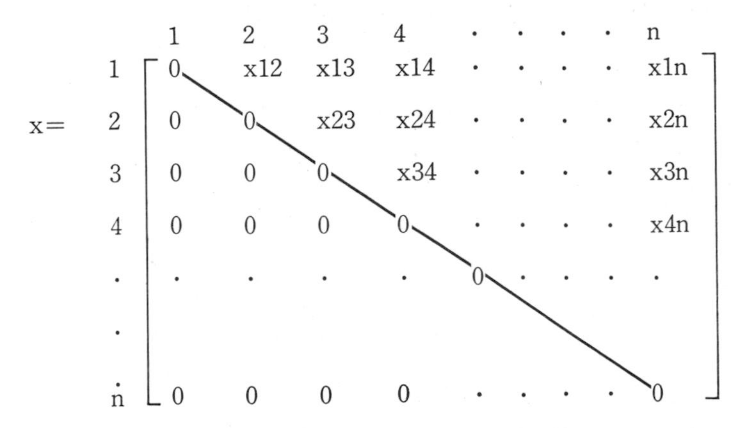

Fig. 1. Matrix of one-link connections among n scientific papers. The rows represent the papers cited, and the columns the source papers.

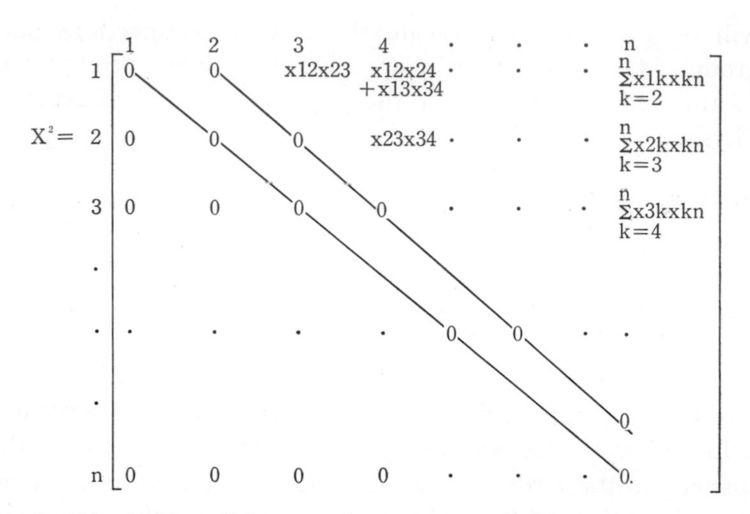

Fig. 2. Matrix of two-link connections among *n* scientific papers.

entire half of elements below the main diagonal is zero because a paper cites only papers which have been previously published; the elements on the main diagonal line are also zero because individual papers do not cite themselves.

Multiplying the original matrix, X, by itself, one can determine how papers in the group are indirectly linked. This squared matrix, X^2, is called a matrix of two-link connections, which indicates two-step indirect links among *n* scientific papers. The elements in the *i* row and *j* column of X^2 is generally shown as $\sum_{k=1}^{n} X_{ik}X_{kj}$. Figure 2 is the matrix X^2 of two-link connections. In this matrix, the elements on the diagonal line next to the main diagonal are also zero.

The element in the first row and third column of X^2 is equal to $X_{12}X_{23}$. If $X_{12}X_{23}=1$, paper No. 1 is said to have a two-link connection with paper No. 3, i.e., paper No. 3 cites paper No. 2, and paper No. 2 cites paper No. 1. This will be denoted by $1 \rightarrow 2 \rightarrow 3$. (Note that paper No. 3 need not cite paper No. 1 directly). If $X_{12}X_{23}=0$, paper No. 1 is said to have no two-link connection with paper No. 3.

The element in the first row and fourth column of X^2 is

equal to $X_{12}X_{24}+X_{13}X_{34}$*. If $X_{12}X_{24}+X_{13}X_{34}=2$, there are two two-link connections between paper No. 1 and paper No. 4, i.e., both $1\rightarrow2\rightarrow4$, and $1\rightarrow3\rightarrow4$. If $X_{12}X_{24}+X_{13}X_{34}+a$, there is one two-link connection between paper No. 1 and paper No. 4, i.e., either $1\rightarrow2\rightarrow4$, or $1\rightarrow3\rightarrow4$. If $X_{12}X_{24}+X_{13}X_{34}=0$, there is no two-link connection between paper No. 1 and paper No. 4.

It is apparent that the original matrix, X, may also be raised to higher powers to obtain the three-link, four-link or even more link connections among *n* scientific papers in the group.

* $\sum_{k=1}^{n} X_{1k}X_{k4}=X_{11}X_{14}+X_{12}X_{24}+X_{13}X_{34}+::::+X_{1n}X_{n4}$

$=X_{12}X_{24}+X_{13}X_{34}$, because all other products are zero.

Part II. An application of the type of matrix to the analysis of a group of collected papers by members of the Japanese Group of Nuclear Physics

The collected papers of the Japanese Group of Nuclear Physics will be analyzed as an example in this section. This group of papers on the meson theory is collected in the *Supplement of the Progress of theoretical physics*, No. 1, in 1955 in commemoration of the 20th anniversary of the proposal of the theory by Yukawa Hideki.[4] For this proposal Yukawa won the Nobel Prize for Physics in 1949. The group consists of 34 important papers on the meson theory by members of the Japanese Group of Nuclear Physics during the period of 1935–50. This group of publications will be called Is in the following discussion.

1. Measurements concerning papers in of group Is.

Figure 3 (omitted here) shows the matrix of one-link connections among 34 papers which belong to group Is. These papers are listed along the rows and down the columns by the method in the previous section. The matrix in Fig. 3 is designated as MIs.

There is a triangular area in the right part of MIs where all the elements are zero. This indicates the time dependence of citations among scientific papers; in later papers, earlier papers are not cited.

The sum of elements in MIs is 67, by which one can indicate the strength of internal connections of group Is. In order to compare different groups which do not consist of the same number of papers, it may be useful to take the ratio between the obtained sum of elements, 67, and the maximum possible sum of elements for the special case of the matrix in which each paper is directly connected with all the papers previously published. The maximum sum is 561 for 34 papers because the maximum sum is $^1/_2\ n(n-1)$ for n scientific papers.

In order to construct the matrix of two-link connections among 34 papers which belong to group Is, matrix MIs is multiplied by itself. This squared matrix in Fig. 4 (omitted here) is $(\mathrm{MIs})^2$.

The sum of elements in $(\mathrm{MIs})^2$ is 157, by which one can also indicate the strength of internal connections of group Is. In order to compare different groups which do not consist of the same number of papers, it may be useful to take the ratio between the obtained sum of elements, 157, and the maximum possible sum of elements for the special case of the squared matrix in which each paper has two-step indirect connections (e.g. $1 \rightarrow 2 \rightarrow 3$) with all the papers previous published. The maximum sum is 5,984 for 34 papers because the sum is $1/6\ n(n-1)\ (n-2)$ for n scientific papers.

2. Measurements concerning papers outside group Is.

Here new measurements to indicate the strength of internal connections of group Is will be proposed by the use of papers that are cited by group Is, but do not belong to group Is.

Table 1 is the distribution of citations according to the number of times that each paper is cited by group Is. The first line of the table, for example, is read as follows: One paper, which is outside of group Is, is cited 8 times by papers that are inside of group Is. Therefore, there are 8 such citations.

For convenience, one can divide these papers outside of group Is into the following two groups: group Ii and group

Table 1 Distribution of Papers Outside of Group Is According to the Frequency with which They are Cited by Group Is

No. of times each paper is cited (x)	No. of papers (y)	No. of such citations (xy)
8	1	8
7	1	7
6	1	6
5	2	10
4	15	60
3	10+ 2[a]	36
2	49+ 2[a]	102
1	194+34[a]	228
Total 36	273+38[a]	457

[a]Indicates numbers of papers by Japanese physicists.

Iw. Group Ii consists of 34 papers cited frequently by group Is, but not belonging to group Is: 30 of these papers are cited more than twice, and four of those are cited twice. Group Iw consists of 34 papers cited less frequently by group Is; these 34 papers are a random sample of papers, cited once by group Is, but belonging to neither group Is nor group Ii. Both groups Ii and Iw consist of papers written by foreign physicists, not by Japanese.

Figures 5 and 6 (omitted here) are the matrix of one-link connections among papers of group Ii, and that of group Iw, respectively. Then, the matrix in Fig. 5 will be called MIi, the matrix in Fig. 6 MIw.

The sum of elements in MIi is 54, and that of elements in MIw is 11, while that in elements of MIs is 67 as mentioned. This means that papers which belong to group Is have the strongest one-link connections among these three groups, Is, Ii, and Iw. On the other hand, papers which belong to group Iw have the weakest one-link connections.

By the matrix multiplication, the squared matrix of MIi, $(\text{MIi})^2$, and the squared matrix of MIw, $(\text{MIw})^2$ are made. (Fig. 7 is omitted here.) The sums of elements of these squared matrices are as follows:

$$\text{Sum of elements of } (MIi)^2 = 70$$
$$\text{,, ,, ,, ,, } (MIw)^2 = 0$$
$$\text{while that of ,, ,, } (MIs)^2 = 157$$

These sums show that papers which belong to group Is have the strongest two-link connections among these three groups, Is, Ii, and Iw, and that papers which belong to group Iw have the weakest two-link connections. This order of the strength of two-link connections is the same order as that of one-link connections.

In addition to the values of the sum of elements in MIs and $(MIs)^2$, one may use the values of the sum of elements in MIi, $(MIi)^2$, MIw, and $(MIw)^2$ as measurements to indicate the strength of internal connections of group Is because these two groups, Ii and Iw, are brought out in connection with group Is. The order of these three groups, Is, Ii, and Iw, in the strength of one-link or two-link connections may be used as another measurement.

3. Measurements concerning groups Is, Ii, and Iw.

In the first place, the construction of the matrix of one-link connections between every two of groups Is, Ii, and Iw will be given to indicate the strength of mutual connections.

In matrix MIis, 34 papers, belonging to group Is, are listed

MIis=

Source Papers

Ii \ Is	1	2	3	4	.	.	.	.
1								
2								
3								
4								
.								
.								
.								
.								

Papers cited in source papers

along the rows, and 34 papers, belonging to group Ii, down the columns in chronological order. The row corresponds to the paper being cited, and the column corresponds to the paper which cites it.

MIsi, MIwi, MIiw, MIws, and MIsw are constructed in a similar fashion. (Figs. 8, 9, 10, 11, 12 and 13 are omitted here.) The element in the first row and first column of MIis is one (1), this means that paper No. 1 of group Is cites paper No. 1 of group Ii. The element in the second row and first column of MIis is zero (0), this means that paper No. 1 of group Is does not cite paper No. 2 of group Ii.

By adding the elements of each matrix, the following sums are obtained:

Sum of elements in MIis	=	129
MIsi	=	24
MIwi	=	16
MIiw	=	32
MIws	=	34
MIsw	=	13

Therefore, the sum of elements in MIis is larger than that in MIsi. This means that group Ii is cited by group Is with greater frequency than group Is is cited by group Ii. The sum of elements in MIiw is larger than that in MIwi. This means that group Ii is cited by group Iw with greater frequency than group Iw is cited by group Ii. The sum of elements in MIws is larger than that in MIsw. This means that group Iw is cited by group Is with greater frequency than group Is is cited by group Iw. Values of the sum of the above 6 matrices are arranged as follows: Sum of elements in MIis > MIws > MIiw > MIsi > MIwi > MIsw.

The above analysis suggests the characteristics of group Is that papers in group Is are closely connected with themselves but not completely closed within themselves. It has been shown that papers in group Is are also connected with other papers outside group Is through the citations.

Summary and conclusions

In the previous sections the group of papers of the Japanese Group of Nuclear Physics, which is called group Is, has been analyzed as an example by the matrix analysis.

It is possible for group Is to indicate the internal connections by the following measurements:

1. The matrix of one-link connections of group Is; this matrix is called MIs. The value of the sum of elements in MIs.
2. The matrix of two-link connections in MIs; this matrix is $(MIs)^2$. The value of the sum of elements in $(MIs)^2$.
3. The matrix of one-link connections of papers which are frequently cited by group Is but not do belong to group Is. This group of papers, which is outside of group Is, is called group Ii; the matrix of group Ii is called MIi. The value of the sum of elements in MIi.
4. The matrix of two-link connections of MIi; this matrix is $(MIi)^2$. The value of the sum of elements of $(MIi)^2$.
5. The matrix of one-link connections of papers which are less frequently cited by group Is, but belonging to neither group Is nor group Ii. This group of papers, which is outside group Is, is called group Iw; the matrix of group Iw is called MIw. The value of the sum of elements in MIw.
6. The matrix of two-link connections of MIw; this matrix is $(MIw)^2$. The value of the sum of elements in $(MIw)^2$.

It is possible to indicate the mutual connections between every two of groups Is, Ii, and Iw by matrices MIis, MIsi, MIiw, MIwi, MIws and MIsw. MIis, for example, is constructed by listing papers of group Is along the rows, and papers of group Ii down the columns in chronological order.

Group Is has the characteristics that papers inside group Is are closely connected with each other but are not completely closed within themselves. This has been shown by the existence of groups Ii and Iw, which are outside of group Is.

It is hoped that other groups of scientific papers will be analyzed by the method that has been developed in this study.

Acknowledgements
The research on which this paper is based was carried out at the Department of History of Science and Medicine, Yale University, under the direction of Professor Derek Price, to whom I am indebted for much encouragement and criticism. Thanks are also due to Dr. George Sorger of the Department of Microbiology of that university for kind assistance in improving the understandability and clarity of the manuscript.

References

1. This was suggested at the seminar on Sociology and Politics of Recent Science, Department of the History of Science and Medicine, Yale University, February 1963. See also Derek Price, "Why does science cumulate", *Frontiers of science and philosophy*, Vol. 2, University of Pittsburgh Press (in press).
2. M. Glanzer, and R. Glaser, *Psychological bulletin*, Vol. 56, No. 5, Sept. (1959), 317. Festinger, L., Schachter, S., and Back, K., "Matrix analysis of group structures", in Lazarsfeld, Paul F. and Rosenberg, M., ed. *The language of social research* (1955), pp. 358–367.
3. For general discussions on the measurement on science, see my paper on the subject: Eri Yagi, *Scientific papers of the College of General Education*, University of Tokyo, Vol. 14, 129, (1964).
4. *Supplement* of the *Progress of theoretical physics*, Kyoto, No. 1 (1955), pp. 1–251.

(3) On the Measurement of Science*

Yagi Eri

The art of measuring science developed mainly after World War II. Measurements of science are required for the study of the history of science and for making national research policies. This is because one can show the historical background, present conditions, and future prediction of scientific activity by the use of these measurements.

In part I of this paper a new classification of possible types of measurements of science is proposed. The types include those theoretically possible, and not just those actually used. In part II several governmental documents are catalogued. Then follows a brief discussion of the types of measurements of science actually used for making national research policies.

I. A CLASSIFICATION OF THE MEASUREMENTS OF SCIENCE

It is possible to classify the types of measurements of science by the use of the fundamental variables of scientific activity, namely, research results, scientists, money, social environment, and time. This is quite different from the classification which makes use of the input-output concept of scientific activity.[1]

1. **Measurements concerning research results**

a. Achievements in science itself: e.g., the energy of particle accelerators,[2] and the number of known elements.[3] Time series of these achievements can be taken.

b. Research work: e.g., the number or ratio of scientific papers,[4] discoveries,[5] inventions,[6] and patents.[7]

c. Communication: e.g., the order of the most frequently

---

* Appeared in *Scientific papers of the College of General Education, University of Tokyo,* Vol. 14, No. 1 (1964), pp. 129–138.

cited journals,[8] the percentages of different journals read by scientists,[9] and the number of journals translated into different languages.[10]

Both categories 'b' and 'c' can be shown in time series. The measurements for categories 'b' and 'c' can be considered for different fields (physics, chemistry, . . . , pure or applied), institutions (government, industry, universities, and non-profitable institutions), industries (aircraft, chemical engineering, . . .), countries (U.S.A., U.S.S.R., . . .), and on a world scale.

2. Measurements concerning scientists

A scientist may be defined, as Derek Price once suggested, as a person who publishes a paper in a scientific journal at any time in his life.

a. The number of scientists,[11] and the number of scientists with different educational backgrounds.[12]

b. The percentage of scientists in relation to national or world population, and the percentage of scientists among various professions.[13]

Both categories 'a' and 'b' can be shown in time series. The measurements are considered for 'a' in different fields, institutions, industries, countries, and on a world scale; and for 'b' in different countries and on a world scale.

3. Measurements concerning money

a. Research and development expenditure.[14] (Hereafter, research and development is abbreviated R & D.)

b. The ratio of R & D expenditure for basic and applied sciences.[15]

c. The percentage of R & D expenditure in relation to gross national income or national expenditure.[16]

d. The percentage of participation by government, industry, university, and non-profitable institutions in the gross national R & D expenditure.[17]

e. The distribution of R & D expenditure for different uses, such as equipment, materials, journals, electricity, and office expenses.[18]

f. Salaries of scientists.[19]

These measurements can be shown in time series. Categories 'a', 'b', 'e', and 'f' can be considered for different fields, institutions, industries, countries, and on a world scale. Categories 'c' and 'd' can be considered for different countries.

4. **Measurements concerning social environment**

a. The background of a scientist: e.g., his political, economic, religious, and educational background.[20]

b. The background of the scientist's parents: e.g., their political, economic, religious, and educational backgrounds.[21]

c. The number of institutions: e.g., the number of universities, observatories, and industrial institutions.[22]

These measurements can be shown in time series. Categories 'a' and 'b' are considered for different fields, institutions, industries, countries, and on a world scale. Category 'c' can be considered for different countries and on a world scale.

5. **Measurements concerning time**

a. Lag time: e.g., lag time between an invention and its actual production, and lag time between publication of a paper and its citations.[23]

b. Lead time: e.g., the U.S. lead time in a particular field of science in comparison with the U.S.S.R.[24]

c. Required time for a particular planning process or a project.

These measurements can be shown in time series. Categories 'a' and 'c' are considered for different fields, institutions, industries, countries, and on a world scale. Category 'b' can be considered for different countries.

6. **Measurements concerning some combinations of the above five variables**

Theoretically, there are ten possible combinations if the five variables are taken two at a time.

a. Research results and scientists: e.g., relation between the number of papers published by an author and the number of authors publishing such papers.[25]

b. Research results and money: e.g., relation between research expenditure and the (monetary) value of the inventions derived from research.[26]

c. Research results and social environment: e.g., relation between the number of papers and the authors' educational backgrounds.

d. Research results and time: e.g., relation between the number of papers and their lag time of citations.[27]

e. Scientists and money: e.g., R & D expenditure per scientist.[28]

f. Scientists and social environment: e.g., relation between the number of scientists and the scale of their firms.[29]

g. Scientists and time.

h. Money and social environment: e.g., the percentage of R & D expenditure in relation to national income and the standard of living of the country. R & D expenditure per inhabitant.[30]

i. Money and time: e.g., relation between gross national R & D expenditure and the lead time of the country.

j. Social environment and time: e.g., relation between economic background and the lag time of inventions.

Time series can be taken in these ten categories. A three-dimensional presentation might be useful for these categories. These measurements are mostly considered for different fields, institutions, industries, countries, and on a world scale.

Theoretically, time series can be taken for most categories. These measurements can be considered for different fields, institutions, industries, countries, and on a world scale; but in actual cases the scales mentioned above are not always considered, due to difficulties in collecting data.

II. MEASUREMENTS OF SCIENCE IN GOVERNMENTAL DOCUMENTS

1. **Developed country: the United States**

a. John Steelman, *Science and public policy*, Vol. I, *A program for the nation* (1947). (Figures for the year 1957 in this book are predictions.) In this report, the following measurements of science are mentioned:

Scientists: the number and ratio of scientists who work in government, industry, and universities, 1920–1947–1957, and the annual production of Ph.D.'s in science, 1913–1947–1957.

Money: annual expenditure for R & D, 1920–1947–1957, and the percentage of R & D expenditure in gross national income, 1920–1947–1957. This report proposed a one-percent R & D expenditure for 1957. The distribution of R & D expenditures to different fields, and institutions, and the donors of R & D funds.

b. Preliminary report to AAAS Interim Committee, "Social aspect of science," in *Science*, Vol. 125 (1957), pp. 143–147. This report mentions the following measurements:

Scientists: the number of scientists in 1930 and 1957.

Money: R & D expenditure in 1930 and 1953, and the expenditure ratio of pure and applied sciences; the expenditure ratio of physical and other sciences.

c. President's Science Advisory Committee, "Strengthening American science" (1958). Measurements mentioned:

Money: governmental expenditure for R & D in 1958, the percentage of government funds in relation to gross R & D expenditure, the percentage of basic research expenditure in relation to governmental R & D expenditure, governmental support for a particular field of science (such as meteorology), the percentage of governmental support for a particular industry (such as aircraft), and the average American contribution to basic research in the national budget.

Measurements in all three of the above documents (a, b, and c) are on the U.S. scale.

2. **Semi-developed country: Japan**

Advisory Council for Science and Technology, "Junen-go no kagaku gijutsu" (Science and technology ten years hence), in *Japanese scientific monthly*, Vol. 13 (1960), Nos. 7 & 8, pp. 5–92.

Scientists: the estimated shortage of scientists in Japan between 1960 and 1970.

Money: the percentage of gross national R & D expenditure in relation to gross national income in Japan in 1955, 1956, 1957 and 1958, in comparison with the percentages in the U.S., Britain, West Germany and France. This report proposed a two-percent R & D expenditure in Japan in 1970.

3. **Less-developed countries:**

a. *India.* M. S. Thacker, "Organization and planning of scientific and technological policies," *United Nations conference on the application of science and technology for the benefit of the less-developed areas, Oct. 30, 1962.*

Scientists: the number of students of science in India, such as B. S., M. S., and Ph.D. candidates at the end of the Third Five-Year Plan of India.

b. *Pakistan.* I. H. Usmani, *Science for survival,* Presidential Address delivered at the 15th Annual Science Conference organized by the Pakistan Association for the Advancement of Science, March 19, 1963.

Money: gross national R & D expenditure in Pakistan, the percentage of the R & D expenditure in relation to national income (0.14%). This report proposed a one-percent R & D expenditure. The distribution ratio of governmental expenditure for science among the science departments of six Pakistan universities, the Council for Scientific Industrial Research, and the Pakistan Atomic Commission.

Social environment: the standard of living of Pakistan, and the percentage of world expenditure for armaments in relation to gross world income, which this report regards as waste.

c. *Ghana.* S. T. Quansah, "Formation of research policies

and program," *United Nations conference on the application of science and technology for the benefit of the less-developed areas, Oct. 26, 1962.*

Money: 1961 expenditures for the West African Cocoa Research Institute and the West African Institute for Palm Oil Research, and the percentage of gross national R & D expenditure in relation to Ghana's income (0.19%).

Time: next five years, within which Ghana will have enough scientists.

As we have seen, measurements concerning scientists and money on a national scale are frequently used in these research policy documents on a national level. The percentage of gross national R & D expenditure in relation to national income is the most common measurement in these documents.

Conclusions

All the possible types of measurements of science can be classified by use of the five fundamental variables: research results, scientists, money, social environment, and time. The measurements of science most frequently used for national research policy decisions are those concerning scientists and money. The percentage of gross national R & D expenditure in relation to national income is the most common measurement for policy decisions.

Acknowledgements

This paper is based on my report to the Seminar on Research Policy, Department of History of Science and Medicine, Yale University, in the spring of 1963. The seminar was conducted by Dr. Stevan Dedijer (now at the Institute of Sociology, University of Lund, Sweden), to whom I am indebted for his useful suggestions and encouragement. Thanks are also due all members of the seminar for their critical discussion of my report.

References

1. Fritz Machlup, *The production of knowledge in the U.S.* (Princeton, 1962), p. 179.
2. M. Stanley Livingston, "Early development of particle accelerators," *American journal of physics*, Vol. 27 (1959), pp. 626–629.
3. Derek Price, *Little science, big science* (New York, 1963), Fig. 11.
4. Derek Price, "Quantitative measure of the development of science," *Archives internationales d'histoire des sciences*, No. 14 (1951), pp. 85–93, Figs. 1 & 2. Prof. Price suggests that the oldest of such figures is in Houzeau & Lancaster, *Bibliogr. gen. de l'astron.* (Brussels, 1882), p. 71.
5. William F. Ogburn & Meyer F. Nimkoff, *Sociology* (New York, 1940), Fig. 132.
6. S. Lilly, *Men, machines and history* (London, 1948), figure on relative invention rate.
7. George E. Folk, *Patents and industrial progress* (New York, 1942).
8. Charles H. Brown, *Scientific serials* (Chicago, 1956).
9. *Ibid.*
10. Boris I. Gorokhoff, *Providing U.S. scientists with Soviet scientific information* (National Science Foundation, April 1962), p. 15.
11. John Steelman, *Science and public policy*, Vol. 1 (1947); Naval Research Advisory Committee, *A report to the Secretary of the Navy on basic research in the Navy*, Vol. 2 (1959); and N. Dewitt, *Soviet professional manpower* (National Science Foundation, 1955).
12. Yagi Eri, "Comment on a paper by Professor Watanabe," *Proceedings of the Xth International Congress of the History of Science, 1962*, Vol. 1 (Paris, 1964), pp. 208–210. (See pp. 105; 107, n.6, this book.)
13. Derek Price, *Little science, big science*, Fig. 3.
14. John Steelman, *Science and public policy, federal fund for science* (1947).
15. Chester I. Barnard, "A national science policy," *Scientific American*, Vol. 197 (Nov. 1957), p. 45.
16. Stevan Dedijer, "Measuring the growth of science," *Science*, Vol. 138 (1962), pp. 781–788, Table 1.
17. Barnard, *op. cit.*
18. Ellis Johnson & Herbert Striner, *Research and development resource allocation, and economic growth* (Operation Research Office, John Hopkins Univ., 1960), Fig. 10.

19. David M. Blank & George J. Stigler, *The demand and supply of scientific personnel* (National Bureau of Economic Research, New York, 1957).
20. Robert Merton, "Science, technology and society in 17th century England," *Osiris*, Vol. 4 (1938), pp. 360–632; and Anne Roe, *The making of a scientist* (New York, 1962).
21. Lewis M. Terman, "Scientists and non-scientists in a group of 800 gifted men," *Psychological monographs: general and applied*, Vol. 68 (1954), pp. 1–41, Figs. 1 & 2.
22. Derek Price, "Diseases of science," *Science since Babylon* (New Haven, 1961), p. 115.
23. C. H. Brown, *Scientific serials* (Chicago, 1956).
24. Jess P. Unger, "The reduction of lead time" (mimeograph; Government 260, Seminar on Science and Policy, Harvard Univ., May 1962), Chart 1.
25. Alfred J. Lotka, "The frequency distribution of scientific productivity," *Journal of the Washington Academy of Science*, Vol. 12 (1926), pp. 317–323.
26. John Harsanyi, "The research policy of the firm," *Economic record*, Vol. 30 (1954), pp. 48–60, Figs. 1 & 2.
27. Paul Weiss, "Knowledge: a growth process," *Science*, Vol. 131 (1960), pp. 1716–1719, Figs. 2 & 3.
28. Ernest Rudd, "Methods used in a survey of research and development expenditure in British industry," *Methodology of statistics on research and development* (National Science Foundation 59–36, Dec. 1958), Table 1.
29. *Science and engineering American industry, final report on a 1953–1954 survey* (National Science Foundation 56–16, 1956), Charts 1 & 3.
30. Stevan Dedijer, "Measurement of science in less developed countries," *U.N. conference on the application of science and technology for the benefit of the less developed areas* (Oct. 1962), Fig. II.

(4) Science Education in Tropical Africa After World War II*

Fujita Chie and Yagi Eri

Few studies on the development of science and technology in Africa have been published, excepting some Unesco publications,[1] although the problem of developing countries has been considered internationally important. The purpose of our study is to throw new light on the conditions of science and technology in Africa from an historical point of view. The results of our preliminary investigation on the educational system of tropical African universities is reported and illustrated with several figures in this paper.

Materials used for this study were the *World of learning* (published in 1948, 1952, 1957, 1962 and 1967), Unesco publications, and university bulletins which have been collected by the Research Institute for Higher Education, Hiroshima University.

The scope of our investigation was limited mainly to the area of tropical Africa, as we were most interested in the development of science and technology in developing countries. Only undergraduate education was considered.

Numbers of students and teachers at tropical African universities

The development of tropical African universities is indicated in Figures 1 and 2, made by using the *World of learning*.[2] Figure 1 shows the number of students at tropical African universities at five-year intervals from 1948 to 1967 (here and elsewhere the interval 1948–1952 is an exception). Each university is taken as representative of the country where it is located.

As shown in Figure 1, the number of students at two newer

* Appeared in *Japanese Studies the history of science*, No. 7 (1968), pp. 143–157.

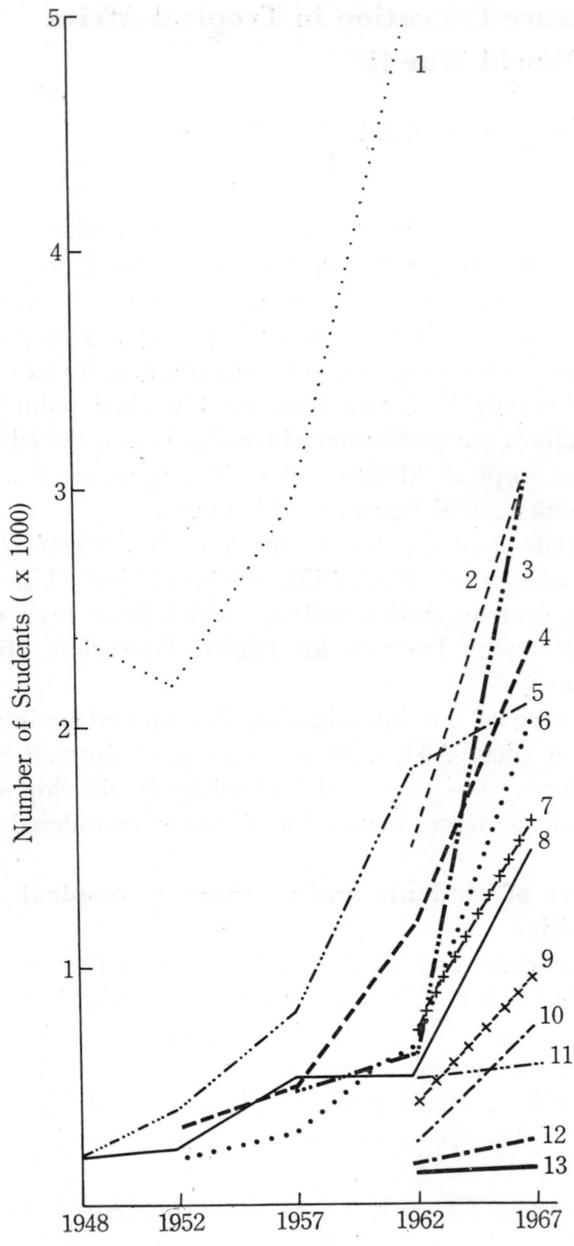

Fig. 1. Number of students at tropical African universities, 1948–1967.

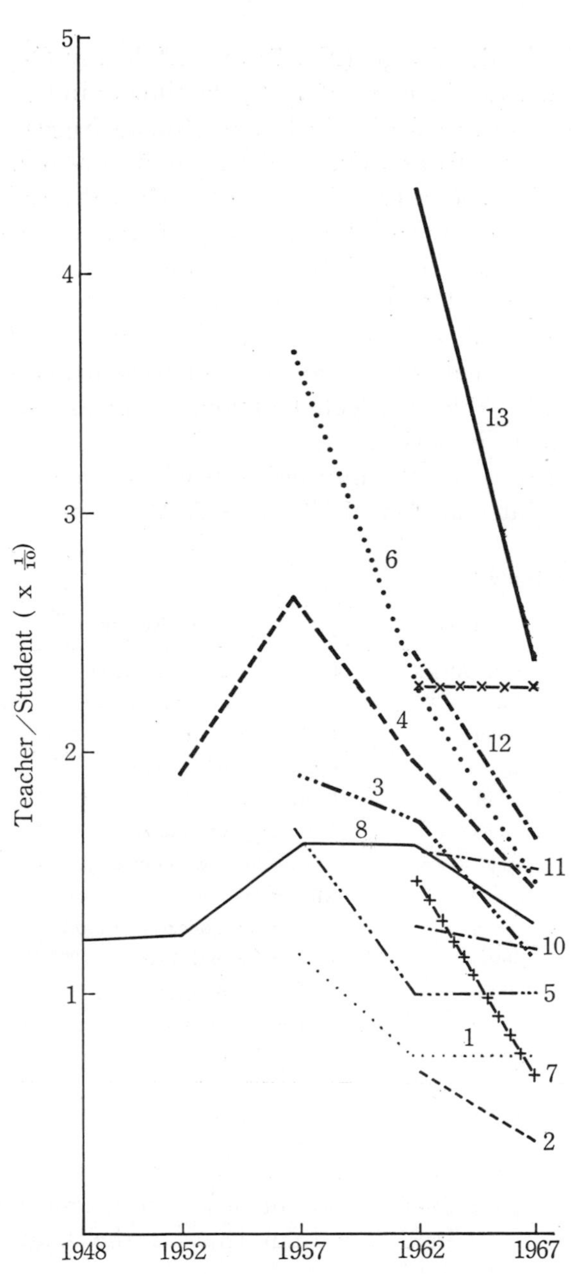

Fig. 2. Number of teachers per ten students in tropical African universities, 1948–1967.

universities in the Congo (No. 7) and Malagasy (No. 2) has rapidly increased since about 1960, the time of independence, and at older universities in the Sudan (No. 5), Nigeria (No. 4) and Ghana (No. 6) as well. Student numbers at some newer universities in Liberia (No. 11), Lesotho (No. 12) and Brunge (No. 13), on the contrary, have not experienced much growth. This retarded progress might be attributed to the less developed social and cultural systems of these countries. One can also note that the size of all tropical African universities is smaller than that of the University of Stellenbosch in South Africa (No. 1), which is selected for comparison as a European-type university in Africa.

Figure 2 indicates the number of teachers per ten students at five-year intervals from 1948 to 1967. Although the number

Key to Figs. 1 and 2

Number	Country	University
No. 1	South Africa	University of Stellenbosch
No. 2	Malagasy	Université de Madagascar
No. 3	Jamaica	University of the West Indies
No. 4	Nigeria	University of Ibadan
No. 5	Sudan	University of Khartoum
No. 6	Ghana	University of Ghana
No. 7	Congo (Leo.)	Université Lovanium de Léopoldville
No. 8	Uganda	Makerere University College
No. 9	Kenya	University College (Nairobi)
No. 10	Libya	University of Libya
No. 11	Liberia	University of Liberia
No. 12	Lesotho	University of Botswana-Lesotho-Swaziland
No. 13	Brunge	Université Officielle de Bujumbra

Source: Reference 2

of teachers in all tropical African universities declined from 1962 to 1967, the teacher-student ratio throughout the period investigated remained less than that at the University of Stellenbosch.

Country of origin of teachers at tropical African universities

To indicate the degree of international influence on the development of tropical African universities, Figure 3 was compiled.[3] It classifies all teachers in the year 1962 by country of

Fig. 3. Number of teachers at tropical African universities, classified by country of origin, in 1962.

Place of origin	A	B	C	Total
England	678	4	20	702
France	5	250	2	257
Belgium	1	132	2	135
U.S.A.	44	4	59	107
U.S.S.R.	1	0	0	1
European Countries (excluding France & Belgium)	93	7	70	170
Asia	54	0	13	67
South Africa	48	0	0	48
Africa (excluding South Africa)	459	42	104	605

Type A: universities originally based on British model.
Type B: universities originally based on French and Belgian models.
Type C: universities not based on any of the above models.
Source: Ref. 3.

origin. In this figure, Type A denotes universities originally based on the British model, Type B those based on the French or Belgian model, and Type C those not based on any of those models. Naturally, most Type A universities are in countries which had been British colonies, and Type B universities are in former French or Belgian colonies. From Figure 3 we can see that the ratio of French and Belgian teachers in Type B universities is higher than that of British teachers in Type A universities. This might be attributable to differences between French and British colonial policies (a more detailed discussion of this matter is beyond the scope of this paper).

Faculties in tropical African universities

The general departmental structure of tropical African uni-

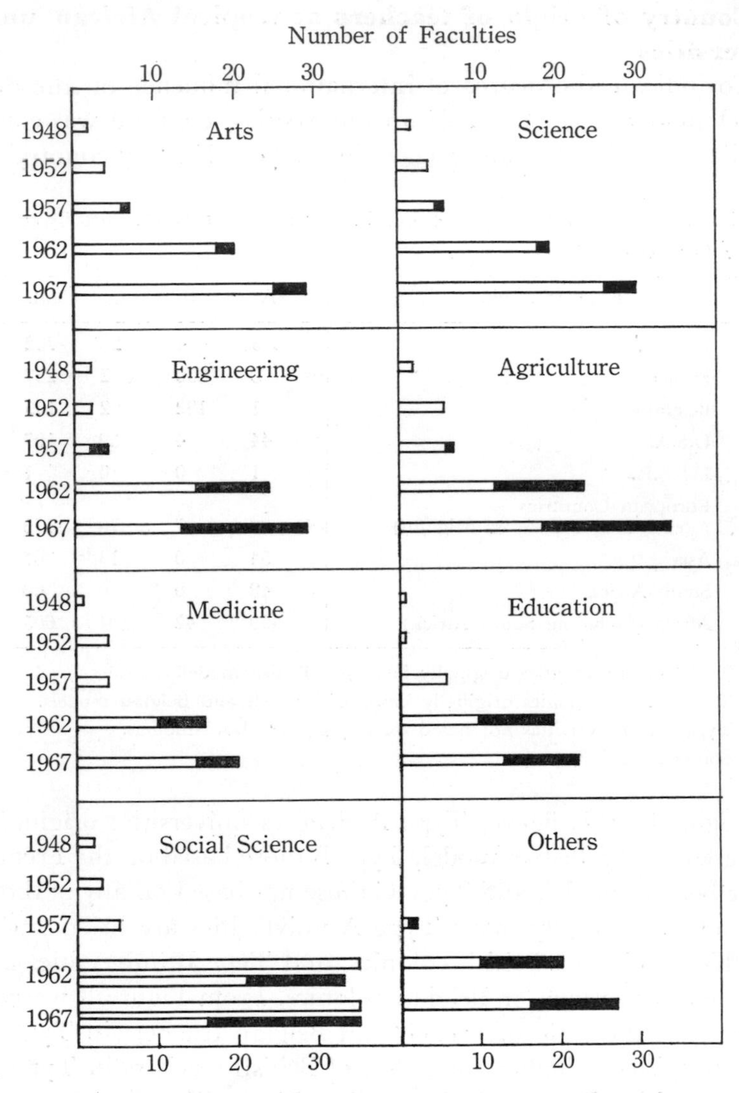

Fig. 4. Faculties in tropical African universities. (Black part of each bar denotes post-secondary institutions.) *Faculty of engineering* includes architecture. *Faculty of agriculture* includes forestry and veterinary science. *Faculty of medicine* includes pharmacy and dentistry. *Faculty of social science* includes law, economics, politics, and commerce. *Faculty of 'others'* includes fine arts, music and African studies.

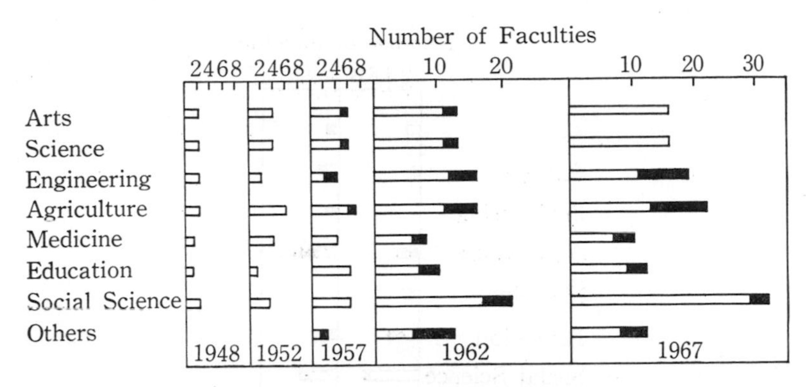

Fig. 5a. Faculties in Type A universities in tropical Africa.

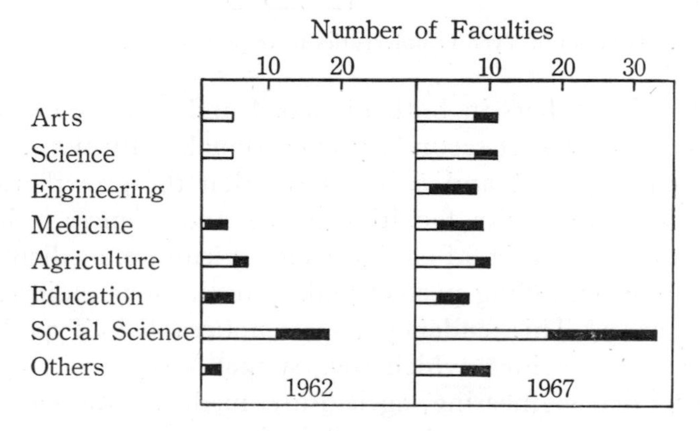

Fig. 5b. Faculties in Type B universities in tropical Africa.

versities at five-year intervals from 1948 to 1967 is shown in Figures 4 and 5.[4] Here the term 'faculty' is used for several kinds of intra-university divisions: faculty, école, institute, and college.

Figure 4 shows the number of various faculties in all tropical African universities during the period studied. Figure 5 has three parts, a, b, and c, which classify the various faculties according to Types A, B, and B. Thus Figure 5a shows faculties in universities originally based on the British model; Figure 5b, those in universities based on the French or Belgian model; and Figure 5c those in universities not based on any of those

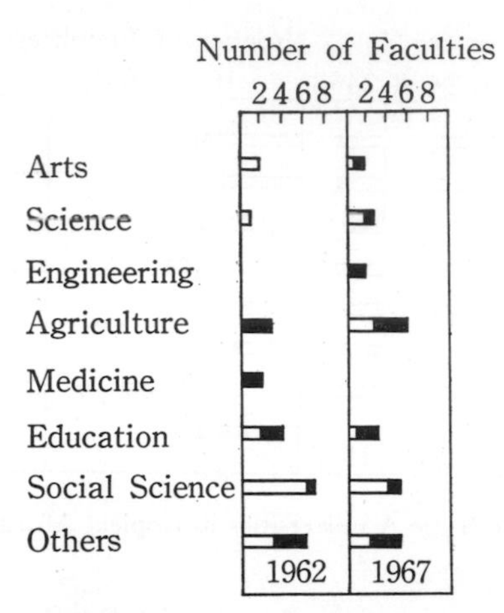

Fig. 5c. Faculties in Type C universities in tropical Africa.

models. Black bars in both Figures 4 and 5 (a, b, c) denote non-university post-secondary educational institutions.

From Figures 4 and 5 it is clear that the growth rate of traditional European faculties of arts and sciences is higher than that of faculties of engineering, agriculture, medicine and education (excluding post-secondary institutions indicated by black bars).[5] This tendency is more noticeable in Type B and Type C universities, which suggest that weakness in applied sciences like engineering, agriculture, medicine and education might prevent the rapid growth of industry in these African countries which urgently need it.

In these applied fields, however, a considerable number of post-secondary institutions have been established in tropical African countries, and some of these institutions have been raised to university status since 1948. Therefore, it is possible that graduates from such institutions will play an important role in the economic and industrial development of their countries. Also, Unesco has a policy of supporting the development of post-secondary institutions in tropical African countries.[6]

Departments of science and engineering in tropical African universities

The departmental structure of tropical African universities is shown in more detail in Figure 6, which is derived from university bulletins.[7] The figure indicates (by a circle) the existence of various departments in eighteen African universities in 1967. This figure also has three parts, 6a, 6b, and 6c. Figure 6a shows departments belonging to the faculty of science, Figure 6b the faculty of engineering, and Figure 6c the faculty of agriculture. Most tropical African universities included in these figures are Type A universities. Eleven of them are tropical African universities, and seven are universities in South Africa, taken for comparison.

Placement of departments in the faculties of science and engineering is generally the same; that is, most tropical African universities commonly have in their science faculties the departments of mathematics, physics, chemistry, zoology, botany, geography and geology; and in their faculties of engineering, the departments of electrical engineering, mechanical engineering, civil engineering, architecture and land surveying.

Apart from the department of land surveying—which seems important to any developing country—the above groupings coincide with those of traditional European universities. This means that the European university system was transferred to universities in tropical Africa with little modification, even after winning political independence. It is often true that scientific organizations tend to be more traditional and conservative than political organizations, due partly to the cumulative nature of science. It does appear, though, that departments in the faculties of agriculture have established closer connections with local African demands than have faculties of science and engineering. (Note that student enrolments in the departments are not treated here.)

Conclusions

Our preliminary investigation of the numbers of students and teachers, and of the departmental structure of tropical African

Fig. 6a. Departments in faculties of science in tropical African universities.

Nation	Department / University	Faculty of Science																									
		①	②	③	④	⑤	⑥	⑦	⑧	⑨	⑩	⑪	⑫	⑬	⑭	⑮	⑯	⑰	⑱	⑲	⑳	㉑	㉒	㉓	㉔	㉕	㉖
Ghana	Ghana	○	○	○	○	○	○	○									○		○	○	○		○				
	Science & Technology		○																				○				○
Nigeria	Ibadan	○	○	○	○	○		○										○									
	Lagos	○	○	○	○	○				○																	
	Nigeria	○	○	○	○	○					○	○						○									
Rhodesia	Rhodesia	○	○	○				○														○	○	○			
Sierra Leone	Fauran Bay	○	○	○	○	○	○	○	○																		
Uganda	Makerere	○	○	○	○	○	○																				
Kenya	Nairobi		○	○	○	○	○	○																	○		
Tanzania	Dar Es Salaam	○	○	○	○	○																					
Congo (Leo.)	Lovanium	○	○	○	○																	○				○	
South Africa	Cape Town	○	○	○	○	○	○				○			○	○												
	Rhodes	○	○	○	○	○	○	○		○		○	○		○	○											
	Witwatersrand	○	○	○	○	○	○	○		○	○				○			○									
	Stellenbosch	○	○	○	○	○		○																			
	Pretoria Technical																										
	South Africa	○	○	○	○	○									○							○					
	North	○	○	○	○	○	○			○			○		○			○						○			

① Mathematics, ② Physics, ③ Chemistry, ④ Zoology, ⑤ Botany, ⑥ Geography, ⑦ Geology, ⑧ Engineering, ⑨ Psychology, ⑩ Microbiology, ⑪ Entomology, ⑫ Pharmacy, ⑬ Astronomy, ⑭ Applied Mathematics, ⑮ Geophysics, ⑯ Oceanography (Fishes), ⑰ Statics, ⑱ Home Science, ⑲ Nutrition, ⑳ Food Science, ㉑ Agriculture, ㉒ Biochemistry, ㉓ Biological Science, ㉔ Meteorology, ㉕ Civil Engineering, ㉖ General Science.

Fig. 6b. Departments in faculties of engineering in tropical African universities.

Nation	Department / University	Faculty of Architecture, Faculty of Engineering																
		①	②	③	④	⑤	⑥	⑦	⑧	⑨	⑩	⑪	⑫	⑬	⑭	⑮	⑯	⑰
Ghana	Ghana Science & Technology	○	○	○	○	○	○		○	○	○				○			
Nigeria	Ibadan																	
	Lagos		○	○	○													
	Nigeria	○	○	○	○			○							○			
Rhodesia	Rhodesia																	
Sierra Leone	Fauran Bay																	
Uganda	Makerere																	
Kenya	Nairobi	○	○	○	○										○	○		
Tanzania	Dar Es Salaam																	
Congo (Leo.)	Lovanium				○													
South Africa	Cape Town		○	○	○			○							○			
	Rhodes																	
	Witwatersrand	○	○	○	○			○				○	○	○				
	Stellenbosch		○	○	○			○	○									
	Pretoria Technical			○	○					○								
	South Africa																○	○
	North																	

① Architecture, ② Electrical Engineering, ③ Mechanical Engineering, ④ Civil Engineering, ⑤ Sanitary Engineering, ⑥ Chemical Engineering, ⑦ Agricultural Engineering, ⑧ Building Engineering, ⑨ Town Engineering, ⑩ Highway Engineering, ⑪ Metallurgy, ⑫ Mining, ⑬ Mining Geology, ⑭ Land Surveying, ⑮ Design, ⑯ Pharmacology, ⑰ General Science.

Fig. 6c. Departments in faculties of agriculture in tropical African universities.

Nation	University \ Department	①	②	③	④	⑤	⑥	⑦	⑧	⑨	⑩	⑪	⑫	⑬	⑭	⑮	⑯	⑰	⑱	⑲	⑳	㉑	㉒	㉓	㉔
		Faculty of Agriculture, Faculty of Veterinary Science																							
Ghana	Ghana	○		○	○								○	○		○	○				○				
	Science & Technology		○	○	○	○															○				
Nigeria	Ibadan		○					○	○									○			○	○			
	Lagos																								
	Nigeria		○				○	○						○			○				○				
Rhodesia	Rhodesia																								
Sierra Leone	Fauran Bay																								
Uganda	Makerere																								
Kenya	Nairobi																								
Tanzania	Dar Es Salaam																								
Congo (Leo.)	Lovanium																								
South Africa	Cape Town																								
	Rhodes																								
	Witwatersrand																								
	Stellenbosch	○	○		○				○	○	○	○			○				○		○		○		○
	Pretoria Technical																				○				
	South Africa																								
	North																								

① Agricultural Economy, ② Agronomy, ③ Crop Production & Husbandry, ④ Horticulture, ⑤ Farm Management, ⑥ Agricultural Mechanization, ⑦ Soil Science, ⑧ Agricultural Chemistry, ⑨ Plant Pathology, ⑩ Genetics, ⑪ Microbiology, ⑫ Botany, ⑬ Zoology, ⑭ Entomology, ⑮ Chemistry, ⑯ Home Science, ⑰ Agriculture, ⑱ Agricultural Biology, ⑲ Agricultural Biochemistry, ⑳ Animal Science, ㉑ Veterinary Clinical Studies, ㉒ Animal Husbandry, ㉓ Veterinary Anatomy & Physiology, ㉔ Forestry.

universities indicates that traditional European faculties of arts and sciences are the most popular ones, and that more attention has been paid to the pure than to the applied sciences.

Criticisms of this "ivory tower attitude" have appeared in Unesco publications.[8] Unesco has emphasized the important role of the university in the social and economic development of African countries, frequently suggesting that the present orientation to traditional faculties and departments should be transformed into one more closely connected with local African needs.[9] Because science and technology often develop according to their own internal factors, it is not entirely certain that education confined to applied science, even if closer to African demands, is necessarily useful for the future development of science and technology in Africa. It is difficult to say categorically that the "ivory tower attitude" plays a wholly negative role.

We are very much interested in the future development of sciences and technology in Africa because Japan itself has been a developing country. An analysis of Japanese data might prove useful for the future planning of science and technology in Africa. One of the present writers has proposed a model for training scientists in a developing country, based on such an analysis,[10] and we would like to continue research activities in tropical African countries.

Acknowledgements

We gratefully acknowledge the help given by African university officials who sent their university bulletins to our research center, the Institute for Higher Education, Hiroshima University. Their continued help is needed for our study. Thanks are also due to members of our research group on higher education for the useful discussion in a series of seminars, supported by a grant from the Ministry of Education of Japan, in 1966 and 1967. We are further indebted to Professor Yokoo Takahide, Director of the Institute for Higher Education, Hiroshima University, and to Mrs. Gomi Ikuko, the United

Nations Depository Library, University of Tokyo, for their kind library services.

References

1. *The development of higher education in Africa* (Unesco, Tananarive conference report, 1962).
Ruth Gruber, ed., *Science in the new nations* (London, 1963).
The teaching of science in African universities (Unesco, Rabat conference, 1964).
Eric Ashby, *Universities: British, Indian, African* (London, 1966).
Thomas R. Odihiambo, "East Africa: science for development," in *Science*, Vol. 158 (1967), pp. 876–881.
Science and technical education in Africa (United Nations, E/CN14/398, 1967).
Report on proceedings of the second consultative meeting between the regional group for Africa of the advisory committee on the application of science and technology to development and the ECA secretariat (United Nations, E/CN14/358, 1966).
2. *The world of learning* (London, 1948, 1952, 1957, 1962, and 1967).
3. *The development of higher education in Africa* (Unesco, Tananarive, 1962), pp. 152–153.
4. Same as ref. 2.
5. History of Science Society of Japan, Nakayama, Kamatani & Yagi, eds., *Nihon kagaku gijutsu shi taikei, kokusai* (Source books on the history of science and technology in Japan, international relations; Tokyo, 1968), pp. 51–98.
6. *The teaching of science in African universities* (Unesco, 1964).
7. The following university bulletins for 1966–1967 were used: University of Cape Town, University of Pretoria, Rhodes University, University of South Africa, University of Stellenbosch, University of the Witwatersrand, University College of Fort Hare, University College of the North, University College for Indians, Natal Technical College, Pretoria Technical College, Université Lovanium de Léopoldville, University of Ghana, University of Science and Technology (Kumasi, Ghana), University College of Cape Coast, Université de Madagascar, University of Ibadan, University of Lagos, Université de Dakar, Fouran Bay College, Centre d'Enseignement Supérieur de Brazzaville, University College of Rhodesia and Nyasaland,

University College Dar Es Salaam, University of East Africa, Makerere University College.

8. *Science and technical education in Africa* (United Nations, E/CN14/398, 1967).
9. *Ibid.*
10. Yagi Eri, *Proceedings of the Xth International Congress of the History of Science, Ithaca & Philadelphia, 1962*, Vol. 1 (1964), pp. 208–210.

Part III

Japanese scientists and their social context

1

Society for the Study of Materialism: Yuiken*

Oka Kunio

I still recall clearly the days when our Society for the Study of Materialism (Yuibutsuron Kenkyūkai, abbreviated Yuiken) was organized. Under the gathering storm clouds of war and mounting fascist oppression, laborers' and farmers' movements were hawkishly suppressed, while the ranks of the unemployed swelled daily. Among the intelligentsia, the questions "what must we do?" and "what *can* we do?" deeply troubled everyone's mind. Only a few of us privately discussed materialism seriously, and some suggested that we should follow the way of the 18th-century French encyclopedists. Two men in the group had already become active in the antiwar movement (*Senmu*) started the year before. Finally we agreed to try cultivating each other's academic interest in materialism, despite its failure thus far to develop in the Japanese cultural and intellectual atmosphere, and for this purpose we organized Yuiken. It was late spring in 1932.

There were seven of us on the organizing committee. Ogura Kinnosuke in Osaka had already published his analysis of "Mathematics in a class society" (in *Shisō*, 1929; translated in part in this volume, pp. 19–23), which had aroused considerable interest in certain intellectual circles. Tosaka Jun also had published before 1929 a series of papers on space theory (in Kyoto University's journal of philosophy), and had just come to Tokyo after having completed his work on the theory of scientific method. I myself had been recently engaged in re-examining the history of science from the viewpoint of Marxism. The three of us might be said to have belonged to the encyclopedists' group. Three others were known as "anti-

* Published in *Nihon no kagakusha* (Scientists of Japan), Vol 5, No. 1 (1970), pp. 31–36; No. 2 (1970), pp. 4–9., and summarized for publication here.

religionists," namely Saigusa Hiroto, a philosopher, and two of his friends, Nagata Hiroshi and Hattori Shisō, who were in the *Senmu* antiwar movement. The seventh was philosopher Honda Kenzo.

The six-man organizing group (Ogura was still in Osaka at the time) began making preparations for the Society in the hot summer of 1932. To begin with, we chose Hasegawa Nyozekan (who died recently; in fact, all of the original organizers except myself are now dead), a well-known journalist, to be chief secretary of Yuiken. Each of us appealed personally to natural scientists to join our Society, and succeeded in recruiting forty as founding members. None of the committee members had official connections with any academic institutions, and naturally we suffered from lack of funds. We always shared equally the expenses, such as for office rent and equipment, yet never even considered installing a telephone. To members outside Tokyo we sent letters, and visited those in Tokyo on foot—our *yukata* (lightweight summer kimono) becoming drenched with perspiration. Despite fascist pressures and our inefficient methods, preparations for the establishment of Yuiken progressed steadily.

The inaugural assembly opened to a full house on a cool autumn day, October 23, 1932, in the Kenchiku Kaikan on the Ginza in Tokyo, followed by our first public lecture meeting, held at the Hoken Kyōkai. It was clear that, even in such difficult times, there were among the intelligentsia quite a number of hidden supporters of our movement. But, from the very beginning, the thought police kept a close watch on our activities, though they had difficulty in understanding our open-minded attitude. We were not afraid of the authorities, since we had no academic status.

The first issue of our organizational journal *Yuibutsuron kenkyū* (Studies in materialism) came out in November of that year (1932). In the lead article on the founding of the Society, chief secretary Hasegawa explained that the Yuiken movement was the second one connected with materialism in Japan, the first being one organized fifty years earlier in the

Meiji period by Nakae Chōmin (1847–1901) for the introduction of French materialism.

Research activities

Yuiken did not have a systematic program of research sections or seminars, nor did its range of activities cover all fields, like politics, economics, history, literature and art. There were, as mentioned above, though, Tosaka's development of unique ideas in philosophy, and Nagata served particularly well as an interpreter of materialism. Besides them, many young members were active in similar work. As for natural sciences, various issues regarding concepts of science and technology were taken up, though the problems often fell within the scope of social science more than natural science.

The second public lecture meeting was held in April, 1933 at Bukkyō Kaikan in the Hongo district of Tokyo. Even before the meeting got started, the hall was surrounded by a number of policemen. As soon as chief secretary Hasegawa announced the beginning of the meeting, the police broke it up, arrested several men in the audience (mostly students) and put them in prison—though they offered no physical resistance to arrest. When this unreasonable suppression was reported in the newspapers the following day, a few members who were well-known natural scientists, such as Terada Torahiko and Koizumi Makoto, announced that they were leaving the Society. Even chief secretary Hasegawa himself shamefully contemplated breaking connections with Yuiken. One of the original committee members, Honda, had already left in February of that year due to illness. Hattori also left after the one-time suppression; and finally Saigusa, who had taken a leading part in the group with Tosaka and myself, severed relations with Yuiken in August, 1933.

Although straightforward suppression occurred only once, and even then on a small scale, the damage inflicted on Yuiken was decisive. This was because our group was only a small, weak private circle, rather than a well-established cultural organization for scientific development. Even so, we were not discouraged, as the remaining members were mostly young students. Our official journal was issued by a professional publishing company for the first two or three months, after which it was managed by the office staff of Yuiken. Its subscribers numbered one thousand, so we could meet our operating expenses with the income from the journal. We held regular research seminars thrice weekly, taking one research section at a time. The editorial activities of the journal were sustained smoothly by cooperative young members, who also took an active part in studying materialism. Those days, in my memory, were the most peaceful and fruitful for Yuiken, and we enjoyed considerable progress in our research. On the other hand, as I look back on my long life as a non-university researcher, the most strenuous periods in my life were the time I spent with Yuiken, and the present.

One of the primary tasks undertaken by Yuiken was the criticism of idealistic philosophy. Though it dominated academic circles in those days, we found philosophical idealism too simple to solve the kind of complicated and serious problems that arose under the fanatic thought control of the 1930's. For example, through such agencies as the Ministry of Education's Institute for Studies in Spiritual Science (Seishin kagaku kenkyūsho), the educational authorities forced on the people all kinds of fascist and semi-feudal ideas. One well-known astronomer actually opposed use of the metric system, even though it had been officially approved in 1886. These are examples of general tendencies we had to counter on the basis of essential, analytic, concrete and systematic criticism. Fortunately, we had the right man for this task—Tosaka Jun, the main leader of Yuiken. He was able to develop a critical standpoint consistent with his own unique ideas, instead of falling back on some ready-made standpoint lifted from standard texts on materialism.

His critical essays were compiled in *Nihon no ideorogii ron* (Theory of Japanese ideology), which was volume two of his collected works together with volume one, *Ideorogii no ronrigaku* (The logic of ideology), and volume three, *Ideorogii gairon*

(Outline of ideology). In the first section of volume two can be found his keen critique of "bibliographical" philosophy and Japanese ideology, and his analysis of reactionism and of the way that cultural controls produce a kind of "Japanism." Tosaka clearly demonstrated the significance of a truly scientific mind that could penetrate the crucial intellectual issues of that time. His assessment of Japanese ideology in volume two, and his papers on the concept of science in volume one, were, I think, the two fundamental pillars of originality by which his philosophy exercised an important influence on various fields.

Other unique areas which Yuiken covered were the analysis of concepts of technology and of the history of technology, though members who devoted themselves to this field were only a few, as were the number of papers produced. We cannot ignore, however, the importance of these efforts because of their impact on our larger scientific movement. Papers concerning these problems published in our journal included Aikawa Haruki's "Concepts of technology and technocracy" (May, 1933), Tosaka Jun's "Social status of engineers" (Nov., 1933), and my own "Systems of labor and technology" (July, 1934) and "On the history of technology" (July, 1937). The last one was a supplement to "An introduction to the study of the history of technology" published in the popular journal *Chūō Kōron.* Tosaka also published a number of papers in various journals. As for books on the subject, Aikawa's *Gijutsuron* (Theory of technology; in the series *Yuibutsuron zensho*, 1936) and Masumoto Setsu's *Gijutsushi* (History of technology) were published. Also, Danilevski's work on the modern history of technology was translated by Masumoto in 1937.

One of our main interests in the history of technology was the definition of 'technology'. Eventually we came to an understanding of technology, from the viewpoint of Marxism, as a system of labor. The same definition had appeared in Bukharin's materialistic interpretation of history, but we preferred to base our definition on Marx's original writings. In his works, Marx made no specific statement that technology means a system of labor. He only said that 'labor' is a simple or complex substance inserted between the worker and the object of his work by the worker himself, to serve as a transmitter of his action to that object. From this premise we developed our own theory of technology as a system of labor, independent of Bukharin. The reason we followed this particular process is that we wanted to define technology theoretically on the basis of concrete material existence, instead of using ordinary concepts. Prior to World War II, however, we could not make any more detailed theoretical progress. It was only after the war that we learned of Diderot's clearcut definition of technology. As Tosaka had already pointed out, one of Yuiken's weaknesses was that we neglected the study of the tradition of French materialism, which might have been an effective tool for criticizing Japanese idealism. It was also after the war that Taketani Mituo and Hoshino Yoshirō proposed their understanding of technology as "the conscious application of the objective laws of nature to the productive process," as against our theory of technology as a "system of labor." [cf. I-3]

The final stage

Yuiken persisted steadily in its unique activities as a small group from 1933 to 1936. We did not have any close contact with other cultural groups, and avoided involvement with other political organizations. Nor did our journal suffer the misfortune during these four years of being prohibited by the government.

However, in July, 1937, when war broke out between Japan and China, a number of antiwar leftists were arrested. They belonged to the laborer-farmer coalition (*Rō-nō*) which, though Marxists, opposed the Japan Communist Party's position. (Most Marxists belonging to the Communist Party were already in prison.) The political climate in which all progressives lived had shifted from a general chill to the deep freeze of mid-winter.

Retaining pride as Japan's encyclopedists, Tosaka and I continued to publish articles in various journals and newspapers, as well as in Yuiken's journal. Direct misfortune descended in January, 1938, when, by an unjust order, the police authorities secretly prohibited all publishers from publishing articles written by Tosaka and me, and by certain left-wing novelists. Under these circumstances, it seemed hopeless to continue our research activities. We held an urgent staff meeting to discuss our options. Opinions were divided between those who insisted on dissolving the Society immediately, and stopping all our activities; and those who agreed with breaking up the Society, stopping the weekly research meetings, and discontinuing the journal, but wanted to begin a new journal in order to salvage at least one function of Yuiken. I myself was the only one who strongly supported the first opinion, while the rest of them, including Tosaka, insisted on the second course. Although discussion continued throughout the night, no agreement could be reached. I came to realize that the whole of Japan was being brought under the strictures of wartime controls, and that, therefore, to publish any kind of journal that lacked a forthright antiwar commitment had no meaning. So, at last, I decided to leave the group, saying goodbye to friends with whom I had studied Marxism for seven years. It was almost morning when I reached this decision. Soon afterward, on November 28, 1938, over one hundred persons, including the former staff and main members of Yuiken from all over Japan, were put in prison. Our group, Yuiken, was thus completely destroyed by suppression.

Never more than a small private circle, Yuiken had many weaknesses, faults and limitations. But in our own way we had made our final protestations against the violent tendencies toward war and fascism, though in reality we accomplished little. At least we had tried to forge from Marxism a philosophical base on which to take our antiwar stand. Lacking some such base, most idealist philosophers and natural scientists, unable to resist the trends toward war and fascism, had collapsed and wound up cooperating with the war. Recalling those events of forty years ago, I am not ashamed of the decisions I made then.

2

The Japanese Research System and the Establishment of the Institute of Physical and Chemical Research*

Itakura Kiyonobu and Yagi Eri

Many people have long considered establishment of the Institute of Physical and Chemical Research a major turning point in the history of Japanese science—but why and how this is so have never been adequately explained. In particular, research to date has not sufficiently clarified the nature and extent of changes in the system of scientific research which marked this event *in relation to* the broader scientific, technological, industrial and economic circumstances of Japan prior to the Institute's establishment.

While some research has been done on the history of Japanese science in the Meiji period (1868–1912), that period was only preparatory to the real beginning of modern science in Japan in the Taisho period (1912–1926), which has been largely neglected. Thus, it is the Taisho beginning of the system of scientific research which must be elaborated if the history of modern science in this country is to be effectively explored.

Japan today is in the midst of a third phase of revolutionary changes in its research system, a phase that ranks with the former two that took place in connection with the Meiji Restoration and World War I. These present changes relate to international incidents like the Bikini bomb tests, as well as to critical domestic issues like the use of atomic energy and the role of the Institute for Nuclear Study, the founding of major interuniversity institutes, and the need to develop scientific and technological systems free from neocolonialist controls. To cope with this new situation we must first analyze what led up

---

* Published in *Kagakushi kenkyū* (Journal of the history of science, Japan), No. 41 (1957), pp. 5–13 ; No. 42 (1957), pp. 22–28.

to it, and this in turn requires a correct understanding of the basis of Japanese science and of its special characteristics—and this task belongs to historians of science.

To be sure, past writers have contributed bits and pieces which, however accurate, do not provide a comprehensive view that is historically convincing. For this we feel it is necessary to begin with a comprehensive analysis of the structure of scientific research at the time when the Institute of Physical and Chemical Research came into being. To discuss this point of departure of modern science in Japan, this essay focusses first of all on certain changes in Japan's scientific research system and its relation to the international background. Secondly, the actual process of the Institute's founding as well as certain tensions inherent in the research system are described. We hope in this way to secure a better basis for future research in the history of modern Japanese science.

Background considerations

To substantiate the claim that the Institute signalled an important turning point in the history of Japanese science, one must review certain important and relevant facts. The initial question is when scientific research facilities or auxiliary agencies first came into existence. Of those concerned with physics and chemistry, one naturally mentions first the faculties of science at the imperial universities, three of which existed by the time World War I began. These were Tokyo, Kyoto and Tohoku universities, established in 1877, 1897 and 1911 respectively. Among government-sponsored research facilities relating to industry, the following may be mentioned: Tokyo Industrial Experiment Station, having connections with the chemical industry and nationalized in 1891; the Ministry of Communications' Electrical Experiment Station, affiliated with the electrical industry, founded in 1891 and nationalized in 1918; and the laboratory of the Transporation Ministry, created in 1891. Another agency supporting research was the Tokyo Academy of Science, founded in 1879; in 1908 it became the Imperial Academy of Science and at that time increased the number of its members related to the natural sciences.

This was the extent of research facilities and supporting agencies concerned with physics and chemistry in Japan before World War I. During and immediately after the war a number of new facilities appeared. Just prior to the establishment of the Institute of Physical and Chemical Research came the Shiomi Physical and Chemical Institute (1916) and the laboratory of the Tokyo Electrical Company (1917–1918), Japan's first industrial laboratory. There were also company laboratories affiliated with Mitsubishi Mining, Mitsubishi Paper Manufacturing, Mitsubishi Shipbuilding, Asahi Glass Company, Furukawa Industries, and Sumitomo Industries. In 1919 the Iron and Steel Research Laboratory was opened at Tohoku Imperial University (from 1922, renamed the Institute for Metals Research). The Aeronautics Institute of Tokyo Imperial University, founded in 1918, was fully nationalized in 1921. Also in 1918, the Electrical Experiment Station was nationalized, and Osaka Industrial Research Institute was started. Then came the Army Research Laboratory (1919) and the Institute of Naval Technology (1923). Japan's first philanthropic foundation concerned with support of scientific research, the Keimei Foundation, was established in 1918 to commemorate the three-hundredth anniversary of Tokugawa Ieyasu's death (1616). Finally, the Ministry of Education in 1918 began for the first time to dispense grants for scientific research.

It is because of this constellation of events that we regard the "era of the Institute of Physical and Chemical Research" as a major watershed in the history of Japanese science.

Development of research facilities in the West

Until the middle of the 19th century scientific research in the West was carried on mainly by scientists in private academies. However, from that period on, support for research increasingly came from newly-founded industries whose existence depended heavily on it. Large-scale laboratories were needed, but were beyond the means of individuals. Universities, or more precisely, the German universities, began to make scientific research the very center of their activities. (Before this time, universities, especially outside Germany, had almost no interest in research in the natural sciences; in fact, they definitely discouraged it.)

> The talented amateurs, who in the earlier part of the century practically monopolized science except for a tiny elite of academicians or, in England, of beneficed clergy and college fellows, gave way to the professors and by the end of the century could no longer compete with them in scientific discovery.
>
> It was otherwise . . . in industrial advance. Even at the end of the century the major innovations were still coming from a race of inventors without a university background, who had learned what little science they needed from books and from their experience in workshops and laboratories equipped with their own hand. The great success of inventors like Edison was, however, to presage a new phase, what might be called the industrialization of invention, with the setting-up of large research labortories. The industrial laboratory and the government research laboratory that came with it brought science back into industry in a new way. The consulting scientist and the scientific entrepreneur were gradually replaced by the whole-time salaried scientist, and the new profession of scientific research worker was created.[1]

In the latter half of the 19th century various universities and manufacturing companies, beginning in Germany, equipped themselves with physical and chemical laboratories. However, the conspicuous successes of inventions deriving from basic research—so evident in the electrical and chemical industries—caused the establishment of research facilities for the direct application of science, larger than those found in most universities.

The first example of this type of facility in Japan was, of course, the Institute of Physical and Chemical Research, modeled after Germany's Imperial Institute for Physics (Physikalisch-Technische Reichsanstalt). As Wilhelm Siemens stated in a letter donating land and money for the construction of the German facility:

> This physical laboratory will prove to be of major benefit to the entire empire, both materially and psychologically. Today the peoples of every nation are engaged in vigorous competition. The victors will be those pioneering in new areas, progressing in new directions and establishing new industries based on science. Looking at such competition throughout history, we find that those who pioneered in new areas, created important new industries or stimulated their expansion almost without exception had to rely on new scientific discoveries. . . . In Germany the inventor's rights are adequately protected by patent laws, while scientific and industrial education are widely diffused through appropriate educational channels. As a result, there are virtually no obstacles to the application of new scientific discoveries in industry. The need is rather for scientific research, for the securing of its foundations. The patronage and support of scientific research, because it enhances the material interests of the country, is consequently of the greatest urgency.[2]

The industrial revolution of the 18th and early 19th centuries, while ultimately based on science, was achieved by people with little if any scientific training as such. Later on in the 19th century, as studies of electro-magnetism began, the electrical industry was founded, and the emergence of chemistry as a new discipline prompted a flourish in the chemical industry. Many new industries that emerged at this time were based on scientific research. Wilhelm Siemens, himself a scientist and an engineer, exploited the 19th-century accomplishments of electro-magnetic research to build the electrical industry. As a capitalist of enormous means, he fully realized the value of research in a capitalist economy then entering its imperialist phase.

Germany's Institute of Physics had originally been projected as the State Institute for Advanced Science and Technology (Obersten Wissenschaftlich Technischen Staatinstitute) fifteen years before it was actually built, after many difficulties, in 1887.[3] It had been proposed to the government authorities in 1872 jointly by the mathematician K. H. Schellbach, Crown Prince Friedrich, Hermann von Helmholtz and Emil du Bois-Reymond. Their proposal, however, was rejected by the Prussian Academy of Sciences. Somewhat later it was submitted to Parliament through the efforts of General Hellmuth von Moltke, head of the Central Geodetic Commission, but made no real progress on this occasion either. Moltke then called a meeting to discuss in detail plans for such an institute, a meeting in which Siemens also participated. In 1883 they drafted a "Proposal for establishment of an institute for the experimental advancement of the natural sciences and precision engineering" but were unable to bring this scheme to fruition. Government authorities were first moved to act, and the institute was finally established, when Siemens, intent on the construction of a properly equipped large-scale facility, finally donated his own money and land for it. These were some of the probes and parries required to make this new agency a reality.

Great Britain and the United States were responding to similar influences when they established in 1899 and 1901 respectively the National Physical Laboratory and the National Bureau of Standards. Germany in 1911 also established the Kaiser Wilhelm Society for the Advancement of the Sciences (Kaiser Wilhelm Gesellschaft zur Förderung der Wissenschaften). With laboratories for physical chemistry, electro-chemistry and general chemistry, this agency played a very important role during World War I.

These laboratories were, of course, designed to meet the demands of industry, but very early on, large industries, especially in the United States and especially those concerned with chemistry and electrical engineering with their own researchers, laboratories and testing facilities, began to expand their activities into the realm of basic research. The first to do so was America's General Electric Company, of which Lawrence A. Hawkins, in his book *Adventure into the unknown: the first fifty years of the General Electric Research Laboratory*, wrote:

> Under such engineering leadership the company's products rapidly grew, in quality, in range of sizes and in diversification. As new applications of electric power developed, they were met by sound engineering practice. So the company grew and prospered. General Electric attained a worldwide reputation as a leader in the electrical industry.
>
> Gradually, however, it was perceived that one thing was lacking. Engineering is applied science. Scientific knowledge is the most important raw material for engineering development and the source of scientific knowledge is research. The engineering progress of the 1890's was based on the researches of Faraday, Joseph Henry, and other scientists in the early part of the century. But unless new research produces new scientific facts, engineering development must inevitably slow down or even come to a halt.
>
> There was little fundamental research in American universities or elsewhere in this country, so why should not General Electric itself engage in such research? Just as the great steel companies find it advantageous to acquire ore fields of their own, why should not General Electric engineering possess a source of its most essential raw material? Why not a G.E. research laboratory? This question became explicitly asked, and was promptly answered by action.
>
> Industrial research was a novelty. There were development laboratories and testing laboratories in existence. Indeed, General Electric had one of the best development laboratories, under Elihu Thomson at Lynn, Massachusetts, but a laboratory having as its prime function research—the seeking of new facts in the electrical field—was a new idea.[3]

Industry-sponsored laboratories for basic research flourished especially in the United States, and new discoveries began to pour forth from such facilities as the Du Pont Company's Eastern Laboratory built in 1902, Bell Telephone Laboratory (1904), the Mellon Institute (1906) and the Eastman Kodak

Laboratory (1913). Some of the research in physics and chemistry done in these laboratories won Nobel prizes. The parent firms were quick to provide generous subsidies, facilities and researchers. For example, the first General Electric Institute, originally housed in a two-story wooden structure, in 1914 moved into a new nine-story ferro-concrete building. Research came to be recognized as a major investment.[4]

Research in Japan before the new Institute

Though the conditions of Japanese research around the turn of the century have already been touched upon, we turn now to the specific conditions under which the Institute of Physical and Chemical Research was built.

Tokyo University, established in 1877, was renamed Tokyo Imperial University in 1886. The period of the Sino-Japanese and Russo-Japanese Wars (1894–1895 and 1904–1905, respectively) greatly stimulated Japanese capitalism and also produced Kyoto and Tohoku Imperial Universities. That these universities were the direct offspring of wartime exploits can be seen, for example, in the "Bulletin of Tohoku Imperial University" published in 1907: "Earlier plans existed to build imperial universities not only in Tokyo and Kyoto but in Tohoku as well, thereby making the issue one of nationwide concern. As the nation's wealth grew with victory in the 1904–1905 war, certain people came to view an increase in the number of Japan's universities as urgent. The government was considering several plans at that time, when fortunately Mr. Furukawa Toranosuke announced his intention to donate about one million yen to the government for construction of two universities."[5]

Even earlier some Japanese capitalists who had acquired fishing rights as part of Japan's new imperialistic policy began investing money in universities and in science, all for the development of the empire. These imperial universities were not however, especially suitable places either for professors or for research. In Article 1 of the Imperial University Ordinance of 1886, for instance, the following statement is found: "The purpose of the Imperial University shall be to teach those arts and sciences essential to the nation and to conduct research into unknown areas." The university was essentially an agency for training bureaucrats, not a properly equipped facility for research. Nor should the phrase "research into unknown areas" be thought to have had any special significance, given the formation in 1893 of the chair system (cf. p. 207), which stressed teaching, and the absence of specific budgetary provisions for research.

Not surprisingly, Japanese scientists who were informed about the development of scientific research organizations elsewhere in the world became quite dissatisfied with conditions at home. These scientists buttressed their demands for new research organizations and laboratories with repeated warnings that Japan was not keeping pace with the rapid development of research organizations abroad.[6]

Sakurai Jōji, one of those most active in the movement to establish the Institute of Physical and Chemical Research, had this to say about university laboratories at the time:

> There was no provision in the university budget for research. All we had was a fund to cover expenses for classroom experiments, more or less proportional to the number of students in each class, and some of this money could be used for research purposes.[7]
>
> In the faculty of science of the University of Tokyo in those days, the operating budget was only about 5,000 yen for physics and 4,000 yen for chemistry. This was hardly enough to pay for the instruments, chemicals, books, journals, specimens, charcoal, gas and electricity needed to teach students, much less to carry on research. On rare occasions, however, we would receive some money for new facilities and could request that some part of it be allotted for the necessary instruments.[8]
>
> Because of these circumstances even the most capable professors and associate professors were rarely able to accomplish anything and unavoidably wasted most of their time. Not an insubstantial number of able young scientists

> who entered graduate school hoping to carry on research were forced to change their career plans and leave academic life. . . .[9] Frequently they took positions in provincial colleges far from major academic centers, where inevitably their research abilities atrophied. . . .[10] These capable scientists were then reduced to poverty, which was simultaneously a regrettable circumstance for our academic community and a grave loss for the country.[11]

At this time, it should also be noted, there was no system of research grants from the Ministry of Education, nor were there yet any private research-support organizations such as the Keimei Foundation, the Saito Benevolent Society, the Hattori Foundation, or the Mitsui Benevolent Society, all of which subsequently provided aid for scientific research. Of this total inadequacy of research funds, Nagaoka Hantarō once remarked, "I first studied theoretical physics because experimental work was completely impossible in Japan at that time" and, after the Institute was founded, "the contrast with today's [1933] experimental facilities is utterly beyond imagination."[12]

As already noted, when the Tokyo Academy of Sciences in 1906 became the Imperial Academy of Sciences, the proportion of its members related to the natural sciences rose. This was not originally a research facility and, far from providing financial assistance to members, all it did, in effect, was to grant members over sixty a kind of pension. Members were nearly always employed as university professors and the designation "Member of the Academy" was little more than an honorary title.[13] According to the Academy's Constitution, its purpose was "to foster the growth of learning and to assist education," but "because the Academy's budget was only the meager sum of 10,000 yen, funds needed for its various activities were completely insufficient and it was seldom to do much to further research within the scientific community."[14] Among the publications of the Academy at this time, the following can be mentioned: *Memoirs of the Imperial Academy*, five volumes published in English for several years from 1912; two volumes in Japanese, *Dai-ichibu ronbun shū* (Collected papers of the First Division); and one volume in English of *Memoirs of the Second Division.* These too were discontinued due to "budgetary limitations and other reasons." One of the few activities the Academy did manage to sustain continuously was the awarding of the Imperial Academy prizes, and that only because of money made available for this purpose by the Imperial Family and two or three wealthy people in private life.[15] The sad state of the Academy was described by Nakamura Seiji as that of a body with "no time, no money and no facilities."[16]

Sakurai Jōji aptly depicted the contemporary state of Japanese science this way:

> The dream of imitating everything had prevailed in Japan from the beginning of the Meiji period until World War I, and scarcely anyone outside the academic community, either in government or in private life, realized the importance of scientific research. Thus, the lack of support for research was a major anxiety not only to the universities but to the learned world in general. And despite the growing enthusiasm of academic circles for research, the general trend is to dash cold water on it. . . . Conditions in this country afford little ground for hope that scientific research can be smoothly developed; all we can do is to wait for more favorable times.[17]

Such, in brief, was the research situation in Japan prior to the establishment of the Institute of Physical and Chemical Research. Apart from it, there were only the three aforementioned laboratories for applied research in physics and chemistry: the Tokyo Industrial Experiment Station (1900), the Electrical Experiment Station (1891, nationalized in 1918), and the Transportation Ministry's Technical Laboratory (1910).

Proposals for founding major laboratories

The situation for scientific research in Japan continued to improve as the country became involved in various wars, a trend not entirely comfortable for scientists living in a country so recently converted to imperialism. Many leading scientists like Sakurai Jōji felt there was nothing they could do to forestall this trend, but as time passed they did begin to demand improved research conditions. The policy of national prosperity and military might (*fukoku kyōhei*) depended, they realized, on scientific research. Western examples to support this view included Faraday's basic electrical experiments which had spawned the electrical industry; the German physics professor Ernst Abbe's use of scientific principles to improve the microscope and optical glass had enabled Germany's optical industry to surpass Britain's; and basic chemical research had permitted the new chemical industry to go beyond the use of materials like natural indigo.

After Japan's victory over Russia in 1905, Nakamura Seiji, associate professor of physics at Tokyo Imperial University, wrote in an article in the newspaper *Jiji shinpō* (March, 1908) that while "Our country's territory has expanded and our wealth seems to increase daily," Japan had in fact barely managed to win the war. Moreover,

> In Europe various nations have fought desperately with each other for superiority. As a result of this war, we too are certain to find ourselves in a highly competitive situation and must not relax our vigilance. . . . But to place the country in a position where its armaments are sufficient to prevent any slight by a foreign country will require a major shift in thinking. To carry out such a shift, the national wealth will have to be increased. *Fukoku kyōhei* is therefore the first principle of our nation's existence.[18]

He pointed out that in several Western countries there were already such facilities as the Imperial Institute of Physics in Germany, the National Bureau of Standards in the United States, the National Laboratory in Great Britain, and the Central Institute of Arts and Manufacture in France. Quoting Wilhelm Siemens' words, he lamented the weaknesses of Japan's research structure when compared to those of the Western nations.

So far as can be determined, Nakamura's appeal for the

establishment of an institute was the first of its kind in Japan. But in June of 1913 Takamine Jokichi returned from the United States and expressed this same view of the scientists much more effectively. Takamine was a member of the Tokyo University engineering faculty's first graduating class and Japan's most eminent applied chemist. After studying in Britain he had taken a position in the Ministry of Commerce and Industry where he soon began to expound to senior Ministry officials and to leading financiers like Shibusawa Eiichi and Masuda Takashi the desirability of establishing a company to manufacture chemical fertilizer. He subsequently founded the first such firm in Japan (later called the Great Japan Chemical Fertilizer Company and presently known as Nissan Chemicals) and, of course, already had strong links with officialdom and the financial community. In 1890 he had gone to the United States to develop his patent rights on a fermentation process. Settling there permanently he quickly established the Takamine Chemical Laboratory and made his important discovery of adrenalin. Takamine, therefore, had a very profound understanding of the significance of research laboratories.[19] With this background, he was an unusually effective advocate of the cause of establishing a "National Science Institute."

Takamine had often thought about the importance of such an institute and had even mentioned it to Shibusawa Eiichi and others. But his idea took concrete form as a proposal only after his return to Japan. Takamatsu Toyokichi, director of the Industrial Experiment Station and a national leader in applied chemistry, supported his proposal. On June 23, 1913, under the sponsorship of Shibusawa and others, they invited Agriculture and Commerce Minister Yamamoto Tatsuo, other officials of the same Ministry, Ōkura Kihachirō, Magoshi Kyōhei and a number of other prominent people, some one hundred and fifty in all, to a meeting at the Seiyōken Hall in Tsukiji, Tokyo, to discuss the matter.

Four years before, in May of 1909, Baron Mitsui of Mitsui Industries and Masuda Takashi had attracted public attention when they convened an assembly of about one hundred people, including Takamine, to discuss the same subject.[20] Again, in early 1913, when Agriculture and Commerce Minister Nakakoji Ren expressed his hope that a way could be found to create a base for science and technology in Japan, the meagerness of science-based inventions by Japanese was scored by various newspapers and magazines. It was emphasized that great inventions require a secure base in science.[21]

In his speech at Tsukiji's Seiyōken Hall, Takamine said that while Japanese industry thus far had developed almost entirely by copying European models, it would have to go beyond this imitative phase and create new industries based on new inventions, if the nation wanted to amass great wealth. "Inventive research today must have its base in scientific theory," he asserted, and insisted that a large-scale laboratory for basic research be established.[22]

Takamine claimed that he was making his demand "in response to the progress of the times." He pointed to the outstanding development of the German chemical industry, explaining, for instance, why indigo production in India and Japan had ceased, and stressed the precedents of the Carnegie and Kaiser Wilhelm Institutes. His final argument was that "Japan has to develop a new policy according to which its entrepreneurs and scientists would exploit existing resources in Asia, change them into valuable goods and sell them all over the world."[23] He also urged that 20,000,000 yen, the cost of one of the newest dreadnaught battleships—which would, after all, soon become obsolete—be spent instead on a research institute.[24]

Takamine's proposal was debated and a committee of thirty members was formed to work for realization of the proposal. His scheme, unlike Nakamura's, had some promise of strong financial backing. When, in 1914, a request for subsidy was presented to the Diet, it unfortunately dissolved shortly thereafter. The committee continued to seek a favorable hearing, but World War I suddenly intervened in July and, as Japan also joined the war against Germany, the committee's plans were temporarily shelved.[25]

World War I and the structure of scientific research in the West

Though significant in many ways, the late 19th-century developments in research organization discussed earlier were quite limited in scope. The importance of scientific research was demonstrated clearly and directly to governments and capitalists alike when World War I erupted. Because German submarines harassed enemy ships, poison gas was used, and so forth, this conflict has been called the first scientific war. Isolated in a world where a single international economy was now a fact, most countries encountered tremendous problems in supplying the necessities of life and in sustaining industrial production. World War I, in short, was not simply a military conflict but a broader national, industrial and economic one as well. In Laurence Hawkins' words:

> In 1914, when war broke out in Europe and the British fleet's blockade placed a stranglehold on German commerce, the rest of the world awoke with dismay to a realization of the dependence of its industries on German laboratories. In industrial research, Germany was far ahead of all other nations. Optical glass, dyes, many kinds of drugs and chemical reagents, many special alloys, and some metals like magnesium were produced only in Germany.[26]

As a result of the war, Britain established a "Commission of the Privy Council for Scientific and Industrial Research" in July, 1916, due to her lack of basic scientific research in the optical glass, dyestuffs, pharmaceutical and other industries. On its recommendation, a Department of Scientific and Industrial Research (DSIR) was created in December, 1916, as a government agency. Attached to the Privy Council, this office not only had its own laboratory but also dispensed large sums of money to research cooperatives (consisting of representatives from each of the country's various industries, which in turn established cooperative research laboratories).[27] The DSIR is still the key agency in the administration of British science and technology today.

In the United States also many came to see by 1916 the need

to mobilize every field of science for the solution of various war-caused technical problems, and by order of President Woodrow Wilson the National Research Council was created as the working arm of the National Academy of Sciences. Originally a temporary agency for the development of new weapons based on cooperation between scientists, technicians, capitalists and representatives of learned societies, it later became a permanent body.[28] Military research institutes also were established as a result of lessons learned from the war. In 1920 the British War Ministry and Admiralty created research divisions, and the United States' Naval Laboratory was established in 1923.[29]

World War I called forth for the first time a profound awareness of the relationship between science, war and technology, and all the involved countries began to show serious interest in government support and control of scientific and technical research. That Germany established its League of Patrons for German Science (Stifterverband für die deutsche Wissenschaft) in 1920 soon after its defeat, despite its possession of the Kaiser Wilhelm Society and many excellent laboratories, was a manifestation of this trend.

During World War I the structure of private industrial research also changed. Referring to conditions in the United States, Laurence Hawkins wrote:

> New laboratories had been created during the war, and their number was to multiply again and again. The G. E. laboratory, instead of being one of a few industrial research laboratories, was now one of hundreds and was to become one of thousands in the next twenty years.[30]

And as Edward Weidlein and William Hamor observed:

> The novel things that had to be accomplished through exigent research were in fact successfully done by the chemists and physicists who were summoned from our universities, and since then these scientists have had established places of importance throughout American industry. The scientists of the other countries participating in the war also demonstrated their ability to produce needed results. All these nations

> saw from these achievements the many possibilities of still greater benefits in the knowledge and the experience of their industrial-research men, and during the past thirteen years have given encouragement to the application of science in manufacturing.[31]

World War I and Japan's scientific research structure

World War I also inaugurated a period of new and rapid growth in Japanese industry, stimulating basic changes in the structure of scientific and technical research. When the war began, Japan's trade with other countries was interrupted as imports of medical supplies, industrial raw materials and machinery came to a halt. This naturally created major difficulties for the country's health, industry and military capability, and the manufacturing community in particular fell into a state of panic.

But before long Japanese capitalism began moving toward unprecedented prosperity. Japanese manufacturers were now able to control domestic markets previously dominated by Western capital interests. Not only that, markets both in war-torn European countries and in China and India were opened to Japanese goods. Because production was unable to keep pace with consumption, commodity prices rose sharply as capitalists expanded their profit margins and increased their capital savings enormously.

Reaping generous profits without bothering to improve technology, the capitalists had no interest in developing technology or science. However, when foreign orders stopped, the contradictions in a Japanese capitalist production system too dependent on Europe soon surfaced and the weaknesses of Japanese technology became patently obvious. This is evident in the fact that Japan could make such gains only in wartime, for when peace returned, the country was once again unable to compete with foreign monopoly capitalism. Strong demands were consequently heard for a relatively autonomous scientific and technical capability, similar to the independent position of the military on which Japanese capitalism had relied up to this time.

The result was that a number of leading bureaucrats, together with some capitalists, scientists and technicians, began strongly to advocate self-sufficiency for Japanese industry, science and technology, appealing to both government and private business. Japan itself had entered the phase of monopoly capitalism since the Russo-Japanese War, and her capitalists were now able and willing to share their sizeable profits for investing in the expansion of science and industry—a need they increasingly recognized. For military purposes, the government too had to admit the importance of technology and science. For these reasons the construction of scientific research facilities in Japan made continual progress during the war.

Our principal concern here, the Institute of Physical and Chemical Research, was the first of these facilities. No sooner had the war begun than the Industrial Chemical Society (Kōgyō Kagaku Kai) with Takamatsu Toyokichi as president submitted to government authorities in October 1914 "A statement of ways for encouraging development of the chemical industry," which read in part:

> The Japanese chemical industry is still in its infancy and we are greatly disturbed by its continuing dependence on foreign sources of supply for most of the important chemical products and raw materials. Moreover, due to the war in Europe, importation of these commodities has become impossible and various industries in Japan have been seriously affected. Not a few difficulties have arisen. We must, of course, move toward meeting this emergency and promote the development of the Japanese chemical industry in the future. It is of the greatest urgency that we discuss our future prospects. In this situation we consider it this Society's responsibility to the nation to debate suggestions regarding basic policies for developing a Japanese chemical industry according to the necessary conditions of modern industrial development. . . .[32] The first thing we must do is to institu-

tionalize chemical training and research and then establish investigative agencies.[33]

Later that same month the Ministry of Commerce and Agriculture established a Chemical Industry Study Commission to consider ways of supporting the chemical industry. Leading figures in chemical research and in applied chemistry were named members. These included Takamatsu Toyokichi, professor emeritus of Tokyo Imperial University; Sakurai Jōji, professor in the faculty of science of Tokyo Imperial University; Watanabe Wataru, professor in the faculty of engineering of Tokyo Imperial University; Nakazawa Iwata, principal of the Kyoto Industrial Academy; Tawara Jun, technical expert in the Hygiene Experiment Station; Kozai Yoshinao, professor in the faculty of agriculture of Tokyo Imperial University; Suzuki Umetarō, professor in the faculty of agriculture of Tokyo Imperial University; Nagai Nagayoshi, professor in the faculty of medicine of Tokyo Imperial University; and Ikeda Kikunae, professor in the faculty of science of Tokyo Imperial University. This Study Commission made a number of proposals to the Agriculture and Commerce Minister, including one entitled "On the promotion of the soda industry," (leading to establishment of Asahi Glass Company); another calling for enactment of a "Law for the encouragement of production in dyestuffs and pharmaceuticals" (leading to establishment of the Japan Dyestuffs Corporation, the Naikoku Pharmaceuticals Corporation, and the Toyo Pharmaceuticals Corporation); and a third called "The expansion of the Industrial Experiment Station and the jurisdictional transfer of the Porcelain Experiment Station." All of these recommendations were implemented. At its first meeting the matter of a chemical research laboratory was immediately raised and its establishment suggested. At the second meeting, however, it was argued that an institute for chemical research alone was too narrowly conceived and that physics as well should be added. This proposal was accepted and added to the existing plan for establishing a chemical research institute. In this way creation of the Institute of Physical and Chemical Research began to assume concrete form (details will be discussed later).

This Institute was not the only such facility created during World War I. In August, 1915, a temporary research institute for physics and chemistry was established at Tohoku Imperial University and began its work through donations, under circumstances in which "because of the war and the cessation of imports, the desirability of scientific research for self-sufficiency in certain commodities has been clearly perceived."[34] This facility's establishment was suggested by President Hojo of Tohoku University because researchers and equipment in the various departments of the University were totally insufficient. As Professor Honda Kotarō, professor of physics at Tohoku, stressed the desperate need for larger-scale facilities for his research on steel, Sumitomo Yoshizaemon, a wealthy industrialist, contributed 21,000 yen (7,000 yen for each of three years) to the institute's second section where physical research was concentrated. Thus, in April 1916, this second section was established within the department of physics in the faculty of science of which Honda became the chairman and research director. Also based on Sumitomo contributions of 150,000 yen for construction and 300,000 yen for equipment, the Institute for Iron and Steel Research of the Tohoku Imperial University opened with two professors, five associate professors and nine assistants in May 1919. Finally in August 1922, this facility was nationalized as the Institute for Metals Research and continues to operate today.[35]

Establishment of the Aeronautics Research Institute at Tokyo Imperial University was also a direct result of World War I. Earlier, in 1904 and 1905 at the time of the siege of Port Arthur, the Army had used anchored balloons (devised with the assistance of Tanakadate Aikitsu, a physicist at Tokyo Imperial University) for harbor reconnaissance. Following the Russo-Japanese War a Temporary Research Group on Military Balloons was established with Army and Navy officers, university professors and meteorologists as members, to do research on balloons and other aircraft. Under its auspices, airplanes were first imported into Japan. As air corps were established in both the Army and Navy, basic aeronautical research became necessary.[36]

Then in 1914 World War I broke out and a movement for aeronautical research began, centering on those university professors who were members of the aforementioned Temporary Research Group. As a result of efforts by Yamakawa Kenjirō, president of Tokyo Imperial University, Tanakadate Aikitsu, Yokota Naritoshi and others, a chair of aeronautical science and the affiliated Aeronautics Research Institute were established at Tokyo Imperial University.[37] A bill authorizing creation of this Aeronautics Research Institute was approved by the Diet simultaneously with one for the Institute of Physical and Chemical Research.[38]

In the following year (1919) the Army submitted a proposal for a National Aeronautics Research Institute separate from the university. In 1921, in accord with the wishes of the Army and Navy, a major expansion of this Institute was undertaken and it acquired a semi-independent status.[39]

Military research laboratories apart from this also made their initial appearance at this time. In 1919 the Army Scientific Institute came into being, followed somewhat later (in 1923) by the Navy Technical Institute. Among laboratories supervised by the Cabinet were the Industrial Experiment Station in Tokyo and the Osaka Industrial Experiment Station. Also, the Electrical Experiment Station was nationalized in the same year that the Osaka facility was built (1918).

Research institutes affiliated with private companies also made their first appearance during these years. The pioneer in this group was the research laboratory of the Tokyo Electric Company, presently known as the Toshiba Matsuda Laboratory. Originally an experiment station, in 1913 it was separated from the technical division of the factory and enlarged. In 1918 it was further enlarged and made an institute directly subject to the company's board of directors.[40] The Tokyo Electric Company had previously received subsidies from the

General Electric Company in the United States, with which it had negotiated technical agreements. In 1910 Dr. D.W. Coolidge invented the tungsten filament in a G.E. laboratory, and Japan began importing it. By 1913 the G.E. product had captured over half the local market, earning extraordinary profits. When the Tokyo Electric Company Laboratory then came into existence, as a result of the threat of world war, it was decided that Japanese manufacturers should adopt a policy of self-sufficiency.[41] Rather than importing products from Germany, which the war had made impossible, various products dependent on research now came to be made by applying accumulated technology. Increased profits were more and more dependent on large research facilities. Though technology thus far had merely relied on specifications and blueprints received from abroad, the need now was to base Japanese production on domestic research, especially basic technical research on products manufactured, as much as possible, from Japanese materials.[42] Already in 1917, the preceding year, Mitsubishi Mining Company, Mitsubishi Paper Manufacturing, Mitsubishi Shipbuilding and Asahi Glass Companies had each established research laboratories. And in 1918 Furukawa, Sumitomo and a number of other firms followed suit.[43]

Nor was the founding of new research laboratories the only manifestation of this trend. In 1918 a new "Universities Ordinance" was promulgated that authorized for the first time establishment of private universities, and in 1920 Waseda University established a faculty of science and engineering. Even earlier the number of chairs in the imperial universities had begun to increase sharply. Tokyo Imperial University's faculty of science had begun in 1893 when the chair system was created with some seventeen chairs. This number increased slowly up to the war years, when it rose by between fifty to one hundred percent. The University of Tokyo's faculty of science presently has forty-eight chairs. In 1912 it had only twenty-seven but between 1918 and 1924, a period of seven years, the number of chairs escalated to forty-one and has increased only slightly since that time.[44]

Another important change in the Japanese research structure during this period was the establishment of the first channels for subsidizing research. As already noted, these included the society commemorating the three-hundredth anniversary of Tokugawa Ieyasu's death (1616), the Keimei Foundation (1918), and in the same year, the Ministry of Education's grants for scientific research.

The prewar movement for an institute

When Nakamura Seiji had spoken several years earlier of the necessity of establishing a really large-scale physical institute, he had addressed himself principally to private citizens of means. He stated:

> I am appealing especially to the more affluent, whom I wish to encourage without delay to establish a physical institute in our empire, so as to strengthen the military capability, increase its wealth, and improve the quality of our manufactured goods. Such a project, however, will require an enormous sum of money. In Germany, for instance, annual expenditure for their leading physical laboratory two or three years ago was of the magnitude of 160,000 yen, and in the United States it was 190,000 yen. . . . If a large house catches fire, one cannot extinguish it by carrying water in wooden buckets. One must flood the fire with a large amount of water all at once, using a pump. While the amount of water may be the same, the effect of using a large amount all at once is very different from that of employing smaller amounts gradually. . . . If in the future a decision to build such a laboratory is made, favorable results will be obtained only if large-scale plans are drawn up and a lump sum of money invested which together are sufficient to its satisfactory development. . . . Will this proposal be accepted by the government? In essence, of course, it almost certainly will be. But money is another matter. It is in this connection that I appeal earnestly to those with wealth, in the hope that they will share some of it for this project.[45]

Some years earlier, in 1907, Baron Yasukawa Keiichirō, one

of the wealthiest men in Kyushu, had invested 4,100,000 yen in Japan's first private industrial vocational school (the Meiji Vocational School)[46] and Furukawa Toranosuke had donated over a million yen to Tohoku and Kyushu Imperial Universities.[47] Somewhat later, Shiomi Seiji, who made enormous profits during the war years from chemical manufacturing, invested one million yen, half of his personal fortune, in the Shiomi Institute for Physics and Chemistry. There were also the precedents of the Sumitomo contribution to the Metals Research Institute of Tohoku Imperial University and Hara Rokuro's subsidization of construction costs for the Electrochemical Laboratory. Some financial support for the creation of such laboratories, consequently, could be expected from business circles.

Takamine Jokichi's procedure in calling for the establishment of a "National Science Institute" was also to address his appeal to the business community. "By way of explaining my wishes briefly, let me say that I want an organization spending annually at least 500,000 yen and hopefully 1,000,000 yen. If we have something between ten and twenty million for basic capital, we will certainly be able to build the kind of facility I am talking about," Takamine declared.[48] In making this proposal, Takamine had in mind an institute which would investigate virtually every branch of science.

In response to this proposal to build a National Science Institute capitalized at 20,000,000 yen, however, "various industrialists, after due deliberation, agreed that while such plans might be drawn up, there was no possibility at all of realizing them under present conditions."[49] In fact, "the general response was unfavorable to so grandiose a scheme."[50] "The result was that Takamine scaled down his proposal, settled on a capitalization fund of a few million yen and made it his principal objective to establish a chemical institute to support Japan's backward chemical industry."[51]

Thus, the proposal to build an institute costing as much as one of the newest dreadnaught battleships was scaled down because the financial community was not very enthusiastic.

However, it was decided that eight scientists, Takamatsu, Sakurai, Ikeda, Kozai, Suzuki, Nagai, Tawara and Takayama (director of the Industrial Experiment Station), all leaders in chemical research and applied chemistry, would form the planning committee. The original scheme formulated by this group called for a chemical research laboratory costing about 5,000,000 yen, with 1,000,000 to be raised each year over a five-year period. Then, with the aid of Shibusawa Eiichi and Nakano Buei, both leading industrialists, some thirty people recommended by the scientific and financial communities were named to a study commission and gradually the proposal took on concrete form. It was no easy matter for the study commission to acquire the entire 5,000,000 yen from private sources, however. Many argued that chemical research should not be an exclusively private activity but a national enterprise, and therefore part of the 5,000,000 yen should be a government subsidy. Proponents of the institute under these circumstances began to realize that they could not necessarily expect assistance from the business community.[52] Nevertheless, plans went forward and a special committee of seven members chosen by the larger group began drafting a bill for establishment of a chemical institute, which in March 1914 was presented to the Diet. Because the Diet was dissolved at this time, the bill could not be considered. Committee members tried to influence the government, but war broke out in July and, when Japan also declared war on Germany, all planning for the Institute was necessarily, though temporarily, suspended.[53]

Moves to establish an institute during the war

Actually, however, the war enhanced prospects for the Institute's establishment, as was noted earlier. After the Industrial Chemical Society presented to the government its "Statement on ways for encouraging development of the chemical industry," the government created the Chemical Industry Study Commission under the chairmanship of Kamiyama Mitsunoshin, Vice Minister of Agriculture and Commerce. At its first meeting in November 1914 the Commission acknowledged that development of a chemical industry depended on fostering research. The Commission also decided to recommend to the Minister of Agriculture and Commerce that establishment of a chemical institute was urgently needed to stimulate industry. There was already at this time, of course, a government-sponsored laboratory related to the chemical industry, namely the Industrial Experiment Station, concerning which the Study Commission observed:

> The source of industrial development lies in scientific research. The progress of the chemical industry relies chiefly on chemical research. It often happens that people confuse industrial chemical research with research in chemistry as an academic discipline. The Study Commission has given this matter considerable attention and our view is that industrial development depends primarily on promoting research.[54]

Takamatsu Toyokichi, a member of the Study Commission and director of the Industrial Experiment Station at that time, declared, "The best procedure would be for basic research to be done in a research institute."[55]

The second meeting of the Chemical Industry Study Commission held in March 1915 discussed the draft of a bill for the creation of the "Chemical Institute" to which were attached a justification for the proposal, a summary of the Institute's probable activities and a budget. The draft provided for construction costs of 4,500,000 yen and annual operating expenses of about 160,000 yen. The Commission's chairman said the proposal could be implemented if the government and the public would each contribute funds for equipment.[56] On this occasion also some Commission members expressed the opinion that "industrial development will encounter difficulties if chemical and physical research are separated, so the Institute should pursue research in both fields jointly."[57] Again the draft proposal to create a Chemical Institute was changed, its backers having moved toward establishment of a private "Institute of Physical and Chemical Research."

Here it is helpful to introduce some of the source materials related to the course of events by which certain specific aspects of the proposal assumed concrete form. Documents exist to show what people concerned with the issue were doing at this time, and these shed considerable light not only on the Institute as such but on the organization of scientific research in Japan during this period as well.

The plan for the Institute of Physical and Chemical Research was considered by a special committee of five members (Nagai, Watanabe, Takamatsu, Sakurai and Kozai) separate from the Chemical Industry Study Commission. These committee men not only sought the cooperation of the physicists (Yamakawa Kenjirō, Kikuchi Dairoku and others), but they were also careful to make their proposals consistent with the plans for the Chemical Institute while drawing up their own "Bill for the creation of an Institute of Physical and Chemical Research." (This draft legislation had nine authors: Watanabe, Takamatsu, Nakano, Nagai, Yamakawa, Kozai, Sakurai, Kikuchi and Shibusawa). Their recommendations were forwarded to government officials, leading industrialists and others whose support was anticipated. The main points in their draft were:

1) Abandoning imitation of the advanced countries, this Institute will encourage creative research, produce inventions, contribute to world progress, promote the well-being of mankind and enhance national prestige.

2) The Institute will undertake creative research in physics and chemistry, aiming to further the sound development of industry.

3) In accord with lessons learned from the war in Europe, the Institute sees its urgent responsiblity to help the nation become independent in military supplies hereafter and to acquire self-sufficiency in industrial materials. . . .

The "Summary Statement of the Proposal" said the Institute's establishment should be financed jointly by contributions from public-spirited persons wishing to commemorate the Taisho Emperor's accession to the throne and by government subsidies. The capitalization fund must be over 10,000,000 yen, facilities were to be ready within ten years, and operating expenses were to be met by interest payments received on the principal, contributions from patrons of the Institute, and from other miscellaneous income. Also included in the Summary Statement was a "Policy Summation" concerning research at the Institute, and this read as follows:

1) Original research in physics and chemistry shall be the principal objective of the Institute of Physical and Chemical Research. By investigating unknown areas of science it will endeavor to harmonize theory with application and relate both to human concerns. In this way the Institute will be able not only to contribute to world progress but to open new areas of knowledge, produce inventions and thereby enhance national prosperity and the well-being of our people. The Institute will also seek to make Japan independent (of outside suppliers) in certain commodities and make the nation's power and wealth secure.

2) With regard to industrial and agricultural pursuits presently being carried on in this country, the Institute will strive through appropriate research activities to improve and to develop them further. The Institute, moreover, will undertake scientific and technical research pertinent to industries needed by the nation though not yet established in this country, thus helping them to develop.

3) While the Institute of Physical and Chemical Research will carry on scientific research, it will also endeavour to respond to the needs of the state and society, aiming to promote public welfare.[58]

Shortly thereafter, the bill calling for establishment of the Institute was presented to a special session of the Thirty-sixth Diet which met from May 20 to June 10, 1915. The text of this proposal follows:[59]

A Proposal for Establishment of an Institute of Physical and Chemical Research.

The development of a nation's welfare and the progress of civilization, it need scarcely be said, principally depends on scientific research and the results of its applications. Taking account of what the advanced Western nations are doing we find that Great Britain has the National Physical Laboratory, France has the Central School for Arts and Manufactures, America the National Bureau of Standards, and Germany the State Institutes for Advanced Science and Technology and the Kaiser Wilhelm Society. These are in every case either national enterprises or joint public-private undertakings. The Western countries, moreover, also possess numerous facilities supported by wealthy philanthropists. One would have to say that science is the source of Western power today. Through several recent wars the prestige of our Empire has risen and its international position has been greatly enhanced. We have equipped ourselves more or less adequately with the material aspects of civilization, but in the realm of scientific research and its technical applications, we are still doing little more than following or imitating others. It is most unfortunate that when scientists here undertake creative research or seek to develop new inventions they often find necessary facilities and supporting agencies to be lacking. Since the war began to disrupt our foreign commerce, many people have recognized, indeed keenly felt, the necessity to consider ways of attaining independence and self-sufficiency in military equipment, medical supplies and basic industrial materials. The government should immediately select an appropriate model from among those available in the major Western countries, formulate a suitable plan and establish an Institute of Physical and Chemical Research. In doing this it would foster and encourage creative research and inventions among our people, strengthen the foundations of several industries, increase our country's wealth and further civilization. Nor are these the only reasons why a long-range plan of this sort should be devised. We believe that implementation of this proposal would also be an appropriate way to commemorate the accession to the Throne of His Majesty the Present Emperor.[60]

When this draft legislation was presented to the lower House of Representatives, some changes were made in the text. Following the section beginning, "the government should immediately select . . . ," it was decided to substitute the words "formulate an appropriate plan but taking account of the financial situation." The House also decided to remove the final sentence regarding commemoration of the Emperor's accession to the throne.[61] When the bill reached the upper House of Peers its sponsors withdrew it due to a congested parliamentary calendar, but the Budget Committee of the Upper House did approve a resolution expressing its hope that government would "provide facilities for physical, chemical, aeronautical and other kinds of scientific research," and the Peers as a body also accepted this.[62]

"Notwithstanding the fact that this proposal was much more ambitious than the earlier one for the Chemical Institute, government officials and the general public alike were increasingly able to see the need for physical and chemical research."[63] A conference to discuss the matter was held at the residence of the Prime Minister, Count Ōkuma Shigenobu, on June 17. Those attending besides the Prime Minister included Ōura Kanetake, Home Minister; Wakatsuki Reijirō, Finance Minister; Ichiki Kitokuro, Minister of Education; Kono Hironaka, Minister of Agriculture and Commerce; Baron Kikuchi Dairoku, Baron Shibusawa Eiichi, Baron Kondō Tsugushige, Dr. Nagai Nagayoshi, Dr. Sakurai Jōji, Dr. Takamatsu Toyokichi, Dr. Watanabe Wataru, Dr. Kozai Yoshinao, together with Nakano Buei, Toyokawa Ryohei, Ōkura Kihachirō, Inoue Junnosuke and Yasuda Zenjirō from the financial community. Shibusawa briefly discussed what had happened since Takamine made his proposal, after which Sakurai Jōji spoke about the problems of founding research laboratories. He said:

> In the past fifty years our country has developed to a degree without precedent among the nations of the world, and since the Russo-Japanese War we have come to be described as a power of the first rank. All this, however, has been accomplished by merely copying what the advanced nations do. One might even say we have entered the ranks of first-class nations only so far as military capability is concerned. Nor would it, I think, be any exaggeration to say that apart

from our skill at imitation and our military prowess we have little enough to boast about.[64]

In this forthright manner Sakurai stated the reasons why a large-scale physical and chemical laboratory had to be established as a national enterprise and said he hoped the government would grant it adequate support. A number of senior scientists then forcefully explained the kind of international position in which Japanese industry and science had been placed, mentioning the proposal to make the country's industry and science more independent. Others then expressed their views. Everyone agreed completely on the necessity for the laboratory but were far from consensus on how to make it a reality. It was finally decided to create a joint public-private study commission to consider ways of implementing the proposal for the laboratory. A commission of eighteen members was formed, of whom five members, Shibusawa, Nakano, Sakurai, Takamatsu and Kamiyama (Vice Minister of Agriculture and Commerce) had special responsibilities. Following numerous deliberations by the committee, Shibusawa and eleven co-authors formally submitted the commission's proposal to the Prime Minister, the Finance Minister and the Minister of Agriculture and Commerce.[65]

The complete text of this frequently quoted document, dated February 21, 1926, read as follows:

A Proposal to Establish an Institute of Physical and Chemical Research

Wishing to contribute to world civilization, to enhance the status of our nation, to lay the foundations of various industries and to increase the nation's wealth, it is imperative that we encourage creative research in the disciplines of physics and chemistry. The recent war, moreover, has taught us the urgent necessity of independence and self-sufficiency in military supplies and industrial materials and has made us all acutely aware of the need for physical and chemical research. Because these kinds of research facilities have not previously existed in our country, certain public-spirited people are hoping to establish an Institute of Physical and Chemical Research. Such a project will require a large sum of money. And since funds from private backers alone will not be sufficient to achieve the goal, we are earnestly hoping the government will make physical and chemical research a national enterprise and quickly take appropriate steps to promote its development in this way. A detailed budget has been included elsewhere with this proposal.

In March 1916 when the Thirty-seventh Imperial Diet convened, the government introduced a bill for the Institute calling for a subsidy in fiscal year 1916 of 250,000 yen. This proposal won the approval of both houses and the following legislation was publicly announced:[66]

Law Number 16, Article 1. A non-profit public corporation is to be established for the purpose of carrying on physical and chemical research and applying its results to foster the growth of industry. The government is authorized by this legislation to subsidize the previously named corporation for a period of ten years by an amount not exceeding 250,000 yen in any one year. The total amount of the aforementioned subsidies, moreover, shall not exceed 2,000,000 yen. . . .

Article 2. The activities of said non-profit corporation shall be supervised by the Minister of Agriculture and Commerce. . . . .

In this way the Institute of Physical and Chemical Research gained the support of the government and the Diet, taking the first step toward its establishment. Then, at the instigation of Prime Minister Ōkuma, a list of one-hundred eighty-two guarantors—leading entrepreneurs from the six largest prefectures—was prepared. Fifty-five members of this group then formed themselves into a Committee of Guarantors for the Establishment of the Institute of Physical and Chemical Research.[67] On October 5, 1916, thirty of the guarantors were named to an Establishment Commission with Shibusawa as chairman, and Sakurai, Takamatsu, Dan Takuma, Wada Yuji, Ōhashi Shintarō, Shō Seijirō and Nakano Buei as the seven permanent members.

Responsibilities for the actual construction design and research facilities were then assigned. The Physical Section's arrangements were entrusted to Nagaoka Hantarō, professor of physics in Tokyo Imperial University's faculty of science, and to Ōkochi Masatoshi, professor of ordnance in the engineering faculty of the same institution. The Chemistry Section was made the responsibility of Ikeda Kikunae, professor of chemistry in Tokyo Imperial University's faculty of science, and to Inoue Jinkichi, professor of applied chemistry in the faculty of science.[68] Chairman Shibusawa and the permanent members were then assigned to relate the aims of the Institute, the plans for it and the work it would carry on to the various industries. In June 1917 they printed and distributed widely a pamphlet entitled "Industry and the activities of the Institute of Physical and Chemical Research." They also visited a number of businessmen and tried to encourage them to contribute to the project.[69] In the pamphlet the Institute's essential activities were listed as follows:

1) Pure scientific research
2) Basic scientific research and its application
3) Research projects requested by various agencies
4) Liaison with each of the experimental stations and research institutes
5) Training of researchers
6) Voluntary research proposed by outside members
7) Subsidization and commendation of research, the development of sound inventions
8) Publications of the results of research and sponsorship of related public meetings.

The pamphlet went on to say:

This Institute, of course, must further the development of industry, and it is assumed that industrialists making use of its services will receive direct benefits. Because Japan is so far behind the Western nations in physical and chemical research, however, the Institute's objectives will not be easily achieved. Nevertheless, by the scale of its operations and the stability of its foundation the Institute expects to become a major research facility for Japanese industry.[70]

Contributions from businessmen were solicited by Shibusawa Eiichi and Nakano Buei, among others. Hara Rokurō, for instance, had initially planned to create an Electro-Chemical Memorial Institute but upon learning of the new project pledged 300,000 yen to it. Similarly the Mitsui and Iwasaki families each pledged 500,000 yen.[71] Baron Shibusawa, chairman of the Establishment Commission and Permanent Members Baron Dan Takuma, an industrialist with an engineering doctorate, Ōhashi Shintarō and Wada Yuji also made contributions while Count Terauchi Masatake invited a number of influential people from the Tokyo-Yokohama area to his home to solicit contributions. Altogether, these people by March 19, 1917 had promised contributions totalling 2,187,000 yen. Shibusawa, on behalf of the Commission, then applied formally for permission to establish the Institute of Physical and Chemical Research, Incorporated. On the following day its creation was officially authorized by Directive No. 3692 from the Minister of Agriculture and Commerce.

H.I.H. Prince Fushimi was named Chairman of the Institute's Board of Trustees, while Baron Kikuchi Dairoku, a leading figure in Japanese mathematics and scientific education who earlier served as President of Tokyo Imperial University, became the Institute's first Director. On April 26, a ten-year grant from the Imperial Family totalling 1,000,000 yen was received. Thus, in September 1917 Institute-sponsored research began at Tokyo, Kyoto and Tohoku Imperial Universities.

Given these circumstances, the newly-created Institute defined its objectives as follows:

> To promote the development of industry, the Institute of Physical and Chemical Research shall perform pure research in the fields of physics and chemistry as well as research relating to its various possible applications. No industry, whether engaged in manufacturing or agriculture, can de-

> velop properly if it lacks a sound basis in physics and chemistry. In our country especially, given its dense population and paucity of industrial raw materials, science is really the only means by which industrial development and national power can be made to grow. The Institute has thus set forth this important mission as its basic objective.[72]

This "objective" as stated was the ideology of the people who labored to make the Institute a reality and it incorporated beliefs enjoying wide popularity at this time. Even so, one should ask again whether in the history of Japanese science such objectives were subsequently achieved or not.

The Institute's growth and problems

No sooner had the Institute come into being than World War I ended, as did the uniquely favorable conditions for Japanese science. Subsequent conditions, in fact, were highly unfavorable. Goods produced by Western monopoly capital with their high standards of technical excellence, which had temporarily disappeared from Japanese and other Asian markets, once again came flooding in like a great wave. Japan's technically inferior products which had acquired new markets and expanded old ones during the wartime boom simply could not compete, and a succession of manufacturing firms fell into bankruptcy.

During the war the demand for Japanese products, led by exports, increased greatly. The government actively sought to promote this and some eighty dyestuff firms came into existence. But in 1918 when Japan's dyestuff industry had just begun to develop, World War I ended and every country which had perceived the wartime relationship of dyestuffs to the production of chemical and military goods began trying to expand into new markets and thus reentered the Far Eastern region. This trend began with an influx of American dyestuffs and intensified with the dumping of similar German products. Over half the domestic factories were forced to cease production. Even by exploiting tariff barriers, only five or six firms in this field were able to continue operations.[73]

Though this situation had been partially anticipated, it still had a devastating effect on the Institute's financial situation. According to the original plan, 1,000,000 yen was anticipated from the Imperial Family, 2,000,000 yen from the government and 5,000,000 from private contributions. Of this amount construction costs were expected to consume about 1,000,000 yen, leaving 7,000,000 yen for basic capitalization from which annual interest receipts of 400,000 yen would support the laboratory's various activities.[74] However, as might have been foreseen, the expected contributions were not forthcoming from the business community, despite the best efforts of Shibusawa, Nakano and the scientists. A *List of General Contributors* reveals the amounts of money pledged by various people as follows: Mitsui Hachirōemon, 500,000 yen; Iwasaki Koyata, 500,000 yen; Hara Rokurō, 300,000 yen; Furukawa Toranosuke, 100,000 yen; Yasuda Zensaburō, 100,000 yen; Sumitomo Yoshizaemon, 100,000 yen; Kubara Fusanosuke, 100,000 yen; Suzuki Iwajirō, 75,000 yen; Ōkura Kihachirō, 50,000 yen; Takamine Jokichi and Shiobara Matasaku, 50,000 yen; Matsukata Kojirō, 50,000 yen and Yamamoto Tadasaburō, 50,000 yen. The amount of money actually received by the Institute was not quite half the amount pledged, and consequently Prince Fushimi's installation as chairman had to be postponed.[75] Thus did the Institute confront the postwar depression. Despite everything, the Commission members were unable to raise the full amount of money and as late as 1922 the Institute had received only slightly more than 3,000,000 yen.[76]

Worse still, expenditures greatly exceeded what the Commission had budgeted. No appropriate site could be obtained from the government, so 400,000 yen had to be spent for that purpose. A sharp rise in the cost of materials and labor forced numerous changes in the original plan, and the sum required by these exigencies was nearly three times the original estimate, or about 3,000,000 yen. The amount remaining for capital investment was only 2,500,000 yen. With annual operating costs dependent on a six percent rate of return from the principal investment, only 150,000 yen was available to sustain all the Institute's activities. The cost of chemicals and other materials used in experiments also rose to two or three times the amount envisioned in the plan and due to a similar rise in the commodity price index far above what it was when the war began, it was also necessary to increase employees' salaries very substantially. Instead of the projected 400,000 yen for annual operating expenses, the Institute at this point needed at least 1,000,000 yen. Were it, in fact, to carry on as originally planned, bankruptcy could be expected within three years.

Facing these realities, the Institute, despite the great expectations of many, could not avoid making substantial changes in its basic plan. A secret conference decision was made to carry on for ten years on the small amount of money already available, striving to produce favorable results during that period. It was also decided that after 1922, annual operating expenses would be limited to 300,000 yen.[77]

The projected annual budget of the Institute was thus reduced by somewhere between 120,000 and 130,000 yen. This meant that the Institute would now operate on one-third the sum provided in the 1917 plan, and only about one-eighth the amount originally suggested by Takamine. Moreover, its capital resources would be exhausted in ten years. The Institute had to forego its original plan of financing its activities from annual interest payments on the principal investment and, instead, find some means of self-support.

As we have already said, this period which we have designated the "era of the Institute of Physical and Chemical Research" was a time of great expectations for the development of Japanese science and technology. It was a time when voices were heard that Japanese industry, technology and science would rise from the imitative stage of dependence on others to the self-generating stage of independence, and when certain steps were actually taken to achieve this. In fact, however, one cannot say the efforts to make the country's science, industry and technology autonomous were successful.

Up to this time, "the basis for the development of Japanese

capitalism had been the semi-feudal, parasitic system of land ownership in the villages. Such property relationships served only to impoverish agriculture, and this gave the capitalist class an opening. Exploiting the great poverty and conditions of servitude existing in the countryside, the capitalists were able to promote industrial production by paying only very low wages to the poverty-stricken farmers who abandoned the countryside for urban factory jobs. With all these advantages, rapid capitalistic development was guaranteed large profit margins. . . . [Japanese] capitalists ordinarily did not carry out technical improvements themselves but, as competition with foreign capital continued, they were forced to make such improvements and thus began developing research facilities using either their own resources or those of the state. The problem, however, could not be fully solved by these means, so they turned to purchasing patents from the advanced countries, importing machinery and other equipment, and employing foreign technicians."[78]

That other scientific research facilities also were established during these years did not change the situation fundamentally. Perhaps building these technical research facilities did improve conditions somewhat, but not at a fundamental level. Even though the Institute did signify a step beyond earlier conditions, its failure to achieve the scale of activities which the original proposal claimed as justification for its establishment bespeaks the absence of any basic change in conditions.

These facts also constitute a direct indication that Japanese technology was still unable to shed its colonial-style dependence on the West. The subsequent history of the Institute also suggests this.[79]

Though unable to continue operations solely through interest received on its capital investment, the Institute at first acquired some income through the sale of patents. But as mentioned earlier, Japanese capitalists, because of the low wage scale, still had no sustained interest in developing science or improving technology themselves, preferring when necessary to import foreign technology. Their attitude made it impossible for the Institute of Physical and Chemical Research to support itself on income from the sale of patents.

The Institute itself had therefore to engage in manufacturing. In time it set up small factories and incorporated itself in order to support its research activities. As the largest research facility in Japan, the Institute enjoyed a number of important successes, including the establishment of nuclear physics in this country. Its economic base, however, remained weak and it was only supported at all because of the wartime economy which came into being after the Manchurian Incident of 1931. In 1938, for instance, 80.7 percent of its factories' production was military-related.[80] The Institute grew rapidly as World War II expanded, and then collapsed along with Japanese imperialism.[81]

Because of these conditions, applied research performed by the Institute never became integral to Japan's basic industries. The Institute's work was oriented essentially to light industry and certain temporary military requirements.

After World War II the Institute was reorganized as the Institute of Scientific Research, and it secured a new lease on life by manufacturing penicillin and streptomycin. Still, it has not attained anything like the position in Japanese society that the prewar Institute of Physical and Chemical Research had. Even penicillin and streptomycin production began to show a deficit once the Korean armistice took effect, due to postwar over-production and an influx of foreign capital and technology. The production division of the Institute therefore was separated and made a commercial firm. It continues as such to the present time, while the Institute has been able to operate on government support.

From the various problems encountered by the Institute of Physical and Chemical Resarch there arose many questions of great significance in the history of Japanese science. Not least significant is the fact that the dream of the scientists and certain capitalists who had hoped to win autonomy for science, technology and industry was not achieved at this time. Rather, the Institute of Physical and Chemical Research had to incorporate itself and managed to accomplish anything at all only by attaching itself to the forces of a militaristic economy.

References

1. J. D. Bernal, *Science and industry in the nineteenth century* (London, 1953), p. 151.
2. Wihelm Siemens, quoted in Nakamura Seiji, *Seikatsu, kagaku, kyōiku* (Life, science and education; Tokyo, 1938), p. 14–21.
3. Laurence A. Hawkins, *Adventure into the unknown: the first fifty years of the General Electric Research Laboratory* (New York, 1950), pp. 2–3.
4. Edward R. Weidlein and William A. Hamor, *Science in action: a sketch of the value of scientific research in American industries* (New York & London, 1931), p. 40.
5. Tohoku Teikoku Daigaku, ed., *Tohoku Teikoku Daigaku ichiran* (Bulletin of Tohoku Imperial University; Sendai, 1913).
6. Nakamura, *Seikatsu, kagaku, kyōiku,* pp. 10–30; Takamine Jokichi, "Kokumin kagaku kenkyūjo setsuritsu ni tsuite" (On the establishment of the National Scientific Research Institute), quoted in Takamatsu Hakushi Shukuga Denki Kankō Kai, ed., *Kōgaku Hakushi Takamatsu Toyokichi den* (Biography of Takamatsu Toyokichi, doctor of engineering; Tokyo, 1932), pp. 293–301. Cf. also Sakurai Jōji, "Rikagaku kenkyū oyobi hatsumei no shōrei ni taisuru kibō" (Hopes for encouraging scientific research and inventions), in *Tōyō gakugei zasshi* (Journal of Eastern arts and sciences), Vol. 33 (Dec. 1916), pp. 833–846.
7. Sakurai Jōji, *Omoide no kazukazu* (Many memories; Tokyo, 1940), p. 18.
8. Sakurai, ref. 6, p. 843; *Omoide no kazukazu,* p. 112.
9. Sakurai, *Omoide no kazukazu,* p. 19.
10. Sakurai, ref. 6, p. 843; *Omoide no kazukazu,* p. 113.
11. Sakurai, *Omoide no kazukazu,* p. 19.
12. Nagaoka Hantarō, "Kenkyūshitsu gaiken: Rikagaku Kenkyūjo Nagaoka Kenkyūshitsu" (Nagaoka's Laboratory in the Institute of Physical and Chemical Research: a general view), in *Kagaku* (Science), Vol. 3 (1933), p. 31.
13. Nakamura, *Seikatsu, kagaku, kyōiku,* p. 22.

14. Sakurai, *Omoide no kazukazu*, p. 113.
15. *Ibid.*, p. 75.
16. Nakamura, *Seikatsu, kagaku, kyōiku*, p. 22.
17. Sakurai, *Omoide no kazukazu*, pp. 19–20.
18. Nakamura, *Seikatsu, kagaku, kyōiku*, pp. 11–12.
19. Shiobara Matasaku, *Takamine Hakushi* (Dr. Takamine; Tokyo, 1926), p. 68.
20. Wakasugi Yoshigoro, *Hatsumei goju-nen shi* (Fifty-year history of inventions; Tokyo, 1947), p. 41.
21. *Ibid.*, p. 44.
22. Takamine Jōkichi, "Kokumin kagaku kenkyūjo . . . ," in *Kogaku Hakushi Takamatsu Toyokichi den* (ref. 6), p. 294.
23. *Ibid.*, p. 300.
24. *Ibid.*, p. 297.
25. Sakurai Jōji, "Rikagaku Kenkyūjo no setsuritsu ni tsuite" (On the establishment of the Institute of Physical and Chemical Research), in *Tōyō gakugei zasshi* (Journal of Eastern arts and sciences), Vol. 33 (May, 1916), pp. 297–302.
26. Hawkins, *Adventure into the unknown*, p. 59.
27. Kagaku Dōin Kyōkai, ed., *Kagaku gijutsu nenkan* (Yearbook of science and technology; Tokyo, 1942), pp. 100, 618.
28. Bernard Jaffe, *Men of science in America* (New York, 1944), pp. 376, 632.
29. *Kagaku gijutsu nenkan*, pp. 625, 636.
30. Hawkins, *Adventure into the unknown*, p. 64.
31. Weidlein & Hamor, *Science in action*, p. 40.
32. Kōgyō Kagaku Kai, "Kagaku kōgyō no hattatsu shōrei ni kansuru iken sho" (Viewpoint on encouraging development of science and industry), in *Kōgaku Hakushi Takamatsu Toyokichi den*, p. 240.
33. *Ibid.*
34. Tohoku Teikoku Daigaku, ed., *Kinzoku Zairyō Kenkyūjo yōran* (Survey of the Institute of Metals Research; Sendai, 1923), p. 2.
35. *Ibid.*, p. 3.
36. Nakamura Seiji, *Tanakadate Aikitsu Sensei* (Professor Tanakadate Aikitsu; Tokyo, 1943), p. 181. Cf. also Ko Danshaku Yamakawa Sensei Kinen Kai, ed., *Danshaku Yamakawa Sensei den* (Biography of Baron Yamakawa, Professor; Tokyo, 1939), p. 311.
37. *Ibid.*, pp. 197, 314.

38. Cf. "Zappo ran" (Miscellany column) in *Tōyō gakugei zasshi*, Vol. 32 (1915), p. 447.
39. *Danshaku Yamakawa Sensei den*, p. 320.
40. Yasui Masutarō, *Tokyo Denki Kabushiki Kaisha goju-nen shi* (Fifty-year history of Tokyo Electric Copmany; Tokyo, 1940), p. 505. Cf. also Kobayashi Akio, "Matsuda Kenkyūjo" (Matsuda Laboratory), in *Kagaku*, Vol. 20, No. 2 (1950).
41. Kobayashi, "Matsuda Kenkyūjo."
42. *Ibid.*
43. Terashima Masashi, *Sekai taishō Nihon kagakushi nenpyō* (Chronology of the history of science in Japan and the world; Tokyo, 1942), pp. 376–379.
44. Tokyo Teikoku Daigaku, ed., *Tokyo Teikoku Daigaku gojū-nen shi* (Fifty-year history of Tokyo Imperial University; Tokyo, 1932), Vols. 1 & 2.
45. Nakamura, *Seikatsu, kagaku, kyōiku*, p. 27.
46. *Danshaku Yamakawa Sensei den*, p. 161.
47. *Tohoku Teikoku Daigaku ichiran.*
48. Takamine, "Kokumin Kagaku Kenkyūjo . . . ," p. 297.
49. Sakurai Jōji, "Rikagaku Kenkyūjo no setsuritsu," in *Kōgaku Hakushi Takamatsu Toyokichi den*, p. 304.
50. "Kagaku Kenkyūjo setsuritsu shinpō," in *Tokyo keizai zasshi* (Tokyo journal of economics), Vol. 68 (1913), p. 537.
51. Sakurai, ref. 49, p. 304.
52. "Kagaku Kenkyūjo Chōsa Kai" (Investigating group of the Institute of Physical and Chemical Research), in *Tokyo keizai zasshi*, Vol. 68 (1913), p. 1135.
53. Sakurai, ref. 25, p. 298.
54. Suzuki Yōsei, "Takamatsu Hakushi to Rikagaku Kenkyūjo" (Dr. Takamatsu and the Institute of Physical and Chemical Research), in *Kōgaku Hakushi Takamatsu Toyokichi den*, p. 310.
55. *Ibid.*
56. Sakurai, ref. 49, p. 304.
57. Sakurai, ref. 25, p. 298.
58. "Rikagaku Kenkyūjo setsuritsu ni kansuru sōan" (Draft for the establishment of the Institute of Physical and Chemical Research), in *Tōyō gakugei zasshi*, Vol. 32 (July 1915), pp. 435–445.
59. "Rikagaku Kenkyūjo setchi ni kansuru kengi an" (Proposal for founding the Institute of Physical and Chemical Research), *Tōyō gakugei zasshi*, Vol. 32 (July 1915), p. 448

60. *Ibid.*
61. Cf. "Zappō ran" (Miscellany column), in *Tōyō gakugei zasshi,* Vol. 32 (1915), p. 448.
62. *Ibid.*
63. Sakurai, ref. 25, p. 299.
64. Sakurai, "Rikagaku Kenkyūjo setsuritsu mondai ni tsuite" (On the problem of the establishment of the Institute of Physical and Chemical Research), in *Tōyō gakugei zasshi,* Vol. 32 (1915), pp. 583–587.
65. Cf. "Zappō ran" in *Tōyō gakugei zasshi,* Vol. 32, (1915), p. 447.
66. Mizutani Kiyoshi, "Setsuritsu no keika, narabi ni genkyō" (Progress toward establishment, and present condition), in *Rikagaku Kenkyūjo ihō* (Bulletin of the Institute of Physical and Chemical Research), Vol. 1 (1922), pp. 80–88.
67. *Ibid.*
68. Ōkōchi Kinen Kai, ed., *Ōkochi Masatoshi: hito to sono jigyō* (Ōkōchi Masatoshi: the man and his works; Tokyo, 1953), p. 62.
69. Sakurai, *Omoide no kazukazu,* p. 22.
70. "Rikagaku Kenkyūjo no jigyō to sangyōkai" (Activities of the Institute of Physical and Chemical Research and the industrial community), in *Tōyō gakugei zasshi,* Vol. 34 (1917), pp. 545–547.
71. Sakurai, *Omoide no kazukazu,* p. 24; and Mizutani, "Setsuritsu no keika. . . . "
72. *Ōkochi Masatoshi,* p. 65.
73. Kawakita Shinpōsha, ed., *Nihon sangyō bunka hensen shi* (History of change in Japanese industrial civilization; Tokyo, 1954), pp. 202–203.
74. Mizutani, "Setsuritsu no keika . . . ," and Sakurai, *Omoide no kazukazu.*
75. *Ibid.*
76. *Ibid.*
77. *Ibid.*
78. Nihon Kyōsantō Kagaku Gijutsu Bu, ed., "Nihon no kagaku, gijutsu no ketsujo to Kyōsantō no ninmu" (Deficiencies in Japanese science and technology, and the task of the Communist Party), in *Zen'ei* (Vanguard), Vol. 10, No. 11 (Nov. 1946). Cf. Taketani Mituo, *Kagaku to gijutsu* (Science and technology; Tokyo, 1950), p. 221.
79. For a brief description of the subsequent development of the

Institute, see "Riken kara Kaken e" (From the Institute of Physical and Chemical Research to the Institute for Sciencetific Research), in the January & February 1956 issues of *Kagaku Asahi* (Science Asahi).

80. Kaminagakura Shinmin, *Shinkō kontsuerun monogatari* (Tales of newly arisen *Konzern*; Tokyo, 1939), pp. 204–239.
81. "Riken kara Kaken e," in *Kagaku Asahi*, Vol. 16, Nos. 1–2 (Jan.-Feb., 1956), pp. 51–55, 57–65.

3

Social Conditions for Prewar Japanese Research in Nuclear Physics*

Hirosige Tetu

It has been argued that research in nuclear physics in Japan began with the initiative of the late Nishina Yoshio and was guided by the "Copenhagen spirit,"[1] that is, a modern approach to cooperative scientific research. While this argument is not itself objectionable, it is not enough, as it entirely lacks any analysis of the social background which made possible the realization of research in nuclear physics based on such a modern approach. This paper, then, attempts to provide an historical analysis of the relevant social conditions of prewar Japanese research in nuclear physics.

The beginning of nuclear physics in Japan

Research in nuclear physics began in Japan around 1930, shortly after Nishina's return from a period of work in Bohr's Institute in Copenhagen (see Table 1). Having gained a re-

Table 1 Early Chronology of Nuclear Physics in Japan

Year	Chronology of nuclear physics	Related social events
1928	*Dec.*, Nishina returned from Copenhagen.	
1929		Worldwide depression.
1930		Rationalization of Japanese industry began.
1931	*May*, Nishina gave lectures on quantum mechanics at Kyoto University.	
	July, Nishina Laboratory established within the Institute of Physical and Chemical Research.	
1932	Discoveries of neutron and positron.	
	Cockcroft-Walton's experiment.	
	Examination of constructing accelerator, Institute of Physical and Chemical Research.	

* Published in *Japanese studies in the history of science*, No. 2 (1963), pp. 80–93.

(Table 1 cont'd)

Year	Chronology of nuclear physics	Related social events
	Invention of counter-controlled cloud chamber.	
		Sept., Manchurian Incident.
	Nov., Confence on nuclear physics at Tokyo Imperial University.	
	Dec., Establishment of Japan Society for the Promotion of Science.	
1933	Construction of a Van de Graaf generator at the Institute of Physical and Chemical Research.	
	Tomonaga-Sakata investigation of positron generation.	
	Apr., Osaka University's Department of Physics opened lectures.	
1934	Construction of Cockcroft machine at Osaka University.	
	Cosmic ray observation with large chamber at Institute of Physical and Chemical Research.	
	Jan., Japan Society for Promotion of Science formed Subcommittee No. 10 on Cosmic Rays.	
	Nov., Yukawa Hideki proposed the meson theory.	
1935	Beginning of continuous cosmic ray measurement at Institute of Physical and Chemical Research.	
1936		February 26 Incident.
1937	Osaka University and the Institute of Physical and Chemical Research completed construction of cyclotrons.	
	Anderson discovered meson.	
	Detection of meson in cosmic rays by the Institute of Physical and Chemical Research.	
		July, Sino-Japanese War broke out.
1938	Yukawa-Sakata-Kobayashi-Taketani investigation of meson theory began.	

putation from his famous Klein-Nishina formula, Nishina returned home in December 1928 and began his efforts to introduce the new quantum mechanics into Japan. In May 1931 he delivered a course of lectures on quantum mechanics at Kyoto University. Among the audience were Yukawa Hideki,

Tomonaga Shin'ichirō and Sakata Shōichi, who were to become leading theoretical nuclear physicists in Japan. Meantime, Nishina in July 1931 established his own laboratory in the Institute of Physical and Chemical Research (cf. p. 231 f.), where his first steps towards experimental nuclear physics research were made. Cloud chambers and counters were made and a plan for constructing a Van de Graaf generator was examined at the Nishina Laboratory.[2]

In the following year (1932) a nationwide study group on nuclear physics was started by Nishikawa Masaharu and Kiuchi Masazō of Tokyo Imperial University, Kimura Masamichi of Kyoto University, Takahashi Yutaka of Tohoku University, and Nishina Yoshio of the Institute of Physical and Chemical Research. They held a conference on nuclear physics at Tokyo Imperial University on November 16–17, where papers on Dirac's electron theory, the absorption and penetration of light in metals, the scattering and absorption of neutrons, and the like were read.[3] In April 1933, at the annual meeting of the Physico-Mathematical Society of Japan, Nishina and Tomonaga presented a paper on the scattering of neutrons by protons, and Yukawa discussed the problem of the intra-nuclear electron. These were the first original papers on nuclear physics in our scientific community. The Nishina-Tomonaga paper dealt with a conjecture, in connection with the preparation of cosmic ray measurement at the Institute of Physical and Chemical Research, that cosmic rays might consist of neutrons, which had just been discovered in the previous year.[4] Yukawa's paper examined a hypothesis that nucleon-nucleon interaction is due to electron exchange. To this hypothesis Nishina remarked that to assume the existence of a kind of electron which obeys Bose statistics might be helpful. This was the first suggestion toward the meson theory, which was first proposed in November 1934 by Yukawa.[5]

Also in 1933 lectures commenced at the Physics Department of Osaka University (est. 1931), where Yukawa held a position as a reader. At the Institute of Physical and Chemical Research, a Van de Graaf electrostatic generator was built, which attained a potential of 0.7 million volts, though it was rather unstable. In the next year, 1934, Kikuchi moved from the Institute of Physical and Chemical Research to Osaka University and initiated the construction of a Cockcroft-Walton machine, while at the Nishina Laboratory of the Institute of Physical and Chemical Research a series of observations of cosmic rays was performed by means of a cloud chamber. In this same year the Japan Society for the Promotion of Scientific Research organized a Subcommittee (No. 10) on Cosmic Ray research, which is discussed later. A rough statistical picture of the growth of nuclear research is given in Table 2, which shows the total number of papers presented

Table 2 Number of Papers Presented to the Annual Meetings of the Physico-Mathematical Society of Japan, 1932–1943*

Year	Total number of papers	Number of papers related to nuclear physics	Percentage
1932	58	0	0
1933	66	2	3
1934	80	12	15
1935	115	17	15
1936	97	11	11
1937	73	9	12
1938	110	25	23
—	—	—	—
1942	168	42	25
1943	171	40	23
1962	755	137	18

* Data from *Nihon Sūgaku-Butsuri Gakkai shi* (Proceeding of the Physico-Mathematical Society of Japan, 1932–1943) and *Butsuri* (Physics, organ of the Physical Society of Japan, Vol. 17, No. 2, 1962).

to the annual meetings of the Physico-Mathematical Society of Japan from 1932 to 1943, the number of those concerning nuclear physics, and the ratio of the latter to the former. The figures for 1939–1941 are incomplete. Figures for the 1962 annual meeting are also shown. This table indicates the rapid increase from 1934 of both the total number of papers and the proportion of those related to nuclear physics.

From the above considerations we may safely conclude that the years 1931 and 1932 marked the start of nuclear physics in Japan. The timing, however, is not accidental, but rooted in historical necessities. In the first place, these years coincided with the turning point of the history of physics in the 20th century. Having developed quantum theories of atoms and molecules, physicists began directing their investigations toward the atomic nucleus. Epoch-making advances in nuclear physics, such as discoveries of the neutron and positron, Cockcroft and Walton's experiment, and Heisenberg's theory of the structure of the nucleus, were made in the year 1932. Japanese physicists naturally tried to share in this new trend in physics. But this is only one side of the matter. The other side, discussed below, is that the period around 1930 was also an important turning point of the history of science in Japan. As I have explained elsewhere,[6] the modernization of science had its beginning in this very period. The basic social conditions which made possible the developments in Japanese research in nuclear physics were rooted in this historical process.

Modernization of research organization

The great worldwide depression that followed the November 1929 crash of the New York stock exchange dealt a heavy blow to Japan's economy. Japanese capitalism had already been suffering stagnation after the postwar boom ended. Shortly before the depression spread over Japan, the government had undertaken a policy of industrial rationalization to bolster Japan's economic competitive power. The depression served to accelerate this policy. Rationalization was carried out through the Temporary Bureau of Industrial Rationalization (Rinji Sangyō Gōrika Kyoku) under the Ministry of Commerce and Industry. Although the policy emphasized liquidation and incorporation of minor enterprises, lengthening work hours, and scientific management of labor, efforts to expand government research institutes and to attain a high technical level should not be overlooked. For example, remarkable technical advances were made in the spinning and weaving industry and, aided by a yen slump, Japan's cotton goods exports surpassed Britain's in 1933.

Given this social background, some scholars expressed the opinion that the government should give generous financial support to scientific research. They argued that, in order to strengthen economic competitive power, the most urgent necessity was to promote scientific activity on the Western model. On January 14, 1931 a number of leading scholars gathered at the Imperial Academy and formed a committee for promoting a foundation that would distribute funds for scientific research. In December 1932, after nearly two years of campaigning, the Japan Society for the Promotion of Scientific Research (Nihon Gakujutsu Shinkōkai) was established with the support of both government grants and private endowments from economic circles. This body was to become the central agency for promoting the modernization of science in Japan.

From World War I, industrialized countries such as the U.S.A., Britain, Germany and France had followed policies for financing and organizing scientific research. Japan's first important step in that direction was the establishment of the Institute of Physical and Chemical Research in 1917. In 1918 the Ministry of Education instituted a system of grants for research in natural science. But these efforts were curtailed by economic recession.

Turning to the financial basis for scientific research around 1930, our consideration is confined to basic research done in the universities and colleges. In government universities, which were rather few in number, research depended mainly on the working budgets called *Kōza-hi*. The *Kōza*, or professorial "chair," was the unit of Japanese government universities; it consisted of one professor, an assistant professor, and one or more assistants. In the period under consideration, the budget for a chair was about $10,000 per year in today's [1963] currency value. This is roughly twice the present-day budget. The total amount of budgets for chairs in all government universities and colleges amounted to roughly $12,000,000 or one and a half times today's total budget.

These budgets were allocated separately to each chair, and were controlled single-handedly by each professor. Universities in Japan have several faculties, which are divided into departments, and each department in turn consists of a number of chairs. In prewar Japan each chair formed a little kingdom in which the professor exercised jurisdiction over his staff. In this system, communications between and within universities were very imperfect, and professorial monopolies of specific subjects hampered cooperative efforts. Hence, generous funding of chairs did little to promote the modernization or the organization of scientific research.

On the other hand, funds for research programs, distributed independently of university chairs, were ridiculously inade-

Table 3 Income and Grants of the Japan Society for the Promotion of Scientific Research, 1933–1947*

(Unit=1,000 yen)

Year	Income			Expenditures for research	Price index**
	Government allocations	Private endowments	Total		
1933	700	1,807	2,507	493	0.95
1934	700	113	813	604	0.96
1935	700	132	832	657	1.01
1936	730	98	828	700	1.05
1937	800	547	1,347	834	1.15
1938	750	254	1,004	988	1.26
1939	1,000	222	1,222	1,239	1.51
1940	1,300	346	1,646	1,543	1.93
1941	1,700	460	2,159	2,144	2.14
1942	2,000	748	2,748	2,350	2.61
1943	2,000	1,011	3,011	2,483	3.01
1944	3,000	101	3,101	3,001	3.07
1945	3,000	553	3,553	2,583	—
1946	2,000	218	2,218	1,789	43.9
1947	6,000	459	6,459	2,478	111.1
(1961)					(365.9)

* Source: Nihon Gakujutsu Shinkōkai, *Gakujutsu kenkyū no haikei* (Background of scientific research; 1953), p. 197.

** Estimated by the Economic Research Institute, Economic Planning Agency, Japan. Average of 1935–1936=1.00. Official exchange rate at time of writing (1963), $1=¥360.

quate. The Ministry of Education grants, distributed only to professors of natural science and technology, amounted to only some tens of thousands of dollars. The sum of grants made by the Imperial Academy and by private foundations together amounted barely to a hundred thousand dollars. Formation of the Japan Society for the Promotion of Scientific Research brought a distinct change in the direction of financing and organizing cooperative research projects and linking their research to industries. The Society annually disbursed hundreds of thousands of research dollars (Table 3).

The real historical significance of the Society, however, lay not simply in the abundance of its resources, but more in its key functions of expanding and organizing scientific research, and coordinating it with industrial and military goals. The Society's grants were of two kinds; those for personal research, and those for cooperative research projects. In the 1930's cooperative projects rapidly increased (Fig. 1).

The research projects were supervised by a Special Subcommittee composed of members from various universities, institutions, military services and business enterprises, who were responsible for coordinating scientific research with industrial and military affairs. So far as the cooperative research

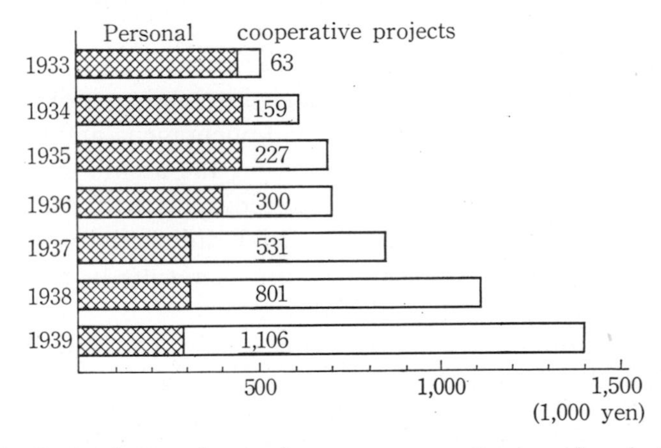

Fig. 1. Ratio of personal research grants to cooperative research project grants made by the Japan Society for the Promotion of Scientific Research, 1933–1939.

was successfully realized, it definitely contributed to raising the level of scientific research in Japan.

Subcommittee No. 10 for Cosmic Rays and Atomic Nucleus

Subcommittee No. 10 was formed on January 29, 1934 for the purpose of investigating cosmic rays. Its chairman was Okada Takematsu, the director of the Central Meteorological Observatory, and the members were Terada Torahiko and Ishimoto Mishio of Tokyo Imperial University, Nishina Yoshio and Kinoshita Masao of the Institute of Physical and Chemical Research, joined shortly after by Nishikawa Masaharu. Effective research was started early in 1935, under the directorship of Nishina. In March 1937 it was reorganized so as to include research on the atomic nucleus, in addition to cosmic rays. Nagaoka Hantarō of the Imperial Academy was appointed the new chairman, and on this occasion Arakatsu Bunsaku of Kyoto University, Kikuchi Seishi of Osaka University, and Sugiura Yoshikatsu of the Institute of Physical and Chemical Research joined the committee, at which time Ishimoto retired. Terada died in December 1935. At the time of the Committee's dissolution in March 1944, Yukawa Hideki of Kyoto University was a member. The stated aims of the committee were:

> . . . to elucidate the nature and the generating mechanism of cosmic rays, to inquire into their interaction with matter, to investigate effects on geophysical phenomena and to study biological effects. The sub-section for the atomic nucleus is to study the structure of the atomic nucleus, the artificial transformation of elements, artificial radioactivity and the like, and to pursue the application of results to medicine, technology, chemistry, and biology.[7]

The committee carried out a number of research projects in pursuit of these aims. Observations and measurements relating to cosmic rays were performed by the members of the Nishina Laboratory in the Institute of Physical and Chemical Research. Measurements of the intensity of cosmic rays were carried out on the summits of Mt. Hakone and Mt. Fuji to investigate the altitude dependence of intensity, in the Shimizu tunnel to see intensity at a great depth, aboard airplanes in flight, and on the vessels Kitano Maru and Heian Maru which sailed to Australia and Seattle respectively to observe latitudinal effects in the Pacific Basin. Continuous measurements with Steinke autographic cosmic ray meters and measurements of the energy of cosmic ray particles by means of a large cloud chamber 60 cm. in diameter and 50 cm. in depth were made. Measurements in the stratosphere were made by means of radiosondes. The main achievements in research on the atomic nucleus were the construction of a 60-inch cyclotron at the Institute of Physical and Chemical Research, and various experiments with the small cyclotrons at the Institute and at Kyoto University on such problems as the production of artificial radio isotopes, biological studies by means of radio isotopes, studies of neutron scattering, and so on.

The activities of Subcommittee No. 10 produced some remarkable results, as described above. Prior to the end of World War II, experimental research in cosmic rays in Japan was the exclusive province of the Nishina Laboratory; and although such research was included in the Laboratory's planning from its beginning, full-scale work was not started until after formation of Subcommittee No. 10.[8] Most of the leading cosmic ray physicists today in Japan—one of the most active countries in this field—were members of the Nishina Laboratory in the days when the Subcommittee's projects were carried on.

The continuous measurement of cosmic rays at the Institute of Physical and Chemical Research, though interrupted before and after the war, ranks high in the world because of its long duration and the certitude of its data. The continuous measurement in Shimizu tunnel, while unfortunately stopped before the end of the war, marked a world record for depth measurement underground. The water-equivalent depth was 3,000 meters. The research group working with a large cloud chamber succeeded in measuring the mass of the meson. It was one of the first determinations of meson mass and the result partially confirmed Yukawa's prediction, thus encouraging the theoretical group of Yukawa, Sakata, Taketani, etc.

The results of cosmic ray research at the Institute of Physical and Chemical Research were also remarkable for their meteorological implications. By analyzing the results of continuous measurement at the Institute, a variation of intensity of cosmic rays according to differences in temperature was shown clearly. Comparing this result with observations carried out on the shipping lines to Australia and Seattle, it was disclosed that the variation of the intensity of a cosmic ray is accompanied by a variation of the temperature of upper air mass.[9] From this disclosure it was decided that from the summer of 1942 the activities of the Nishina Laboratory should concentrate on investigating correlations of cosmic rays and meteorological phenomena, and a cooperative relationship with the Central Meteorological Observatory was initiated.[10] By 1944 the investigations reached the stage where, from observation of cosmic rays on ground level, one could estimate fairly accurately the temperature of the upper air mass.

Research on the atomic nucleus was not so fruitful as cosmic ray research. The principal part of the 6-inch cyclotron at the Institute of Physical and Chemical Research was completed in February 1939, but this machine did not work successfully until the end of 1944, primarily due to the low level of Japanese technology. And shortly after the war, on November 24, 1945, the cyclotrons in Japan, including the Institute's 60-inch one, were destroyed by the U.S. Army. It was not until the end of 1952 that Japanese physicists could again possess a cyclotron. Given this situation, experimental nuclear physics in Japan before and during the war produced no tangible results. Intangible contributions, however, should not be overlooked. For example, prewar experiences, especially at Osaka University, cultivated the ground for today's technical developments in instrumentation and experimentation in nuclear physics, which now is successfully pursued in several institutions, such as the Institute for Nuclear Studies at the University of Tokyo. Moreover, experimental research activity inspired the theorists.

Recalling the prewar days, Sakata once said that active theoretical investigation had been possible only at Osaka University, which at that time was the center for experimental nuclear physics in Japan.[11] Also, the interconnection between the attempt of Nishina, K. Birus, Sekido Yotarō and Miyazaki Yukio at the Physical-Chemical Institute to detect a neutral meson in cosmic rays (1939) and the theory of Sakata and Tanikawa on natural disintegration of the neutral meson (1940)[12] should be noted. In general, theorists before the war had an advantage over the postwar theorists in having constant communications with experimentalists.

To sum up, we may safely say that the work of Subcommittee No. 10 provided a basis for later developments in nuclear physics in Japan.

Social background of Subcommittee No. 10

The funds allocated for Subcommittee No. 10 by the Japan Society for the Promotion of Scientific Research were extraordinarily large for those days. Table 4 shows the main subcommittee and special committee grants in descending order during the period 1933–1940. From this table it is clear that Subcommittee 10's appropriations were surprisingly large, compared with those of other projects, which were mostly related to military needs. It is quite remarkable that research in cosmic rays and nuclear physics occupied so large a part of the total outlay of the Japan Society for the Promotion of Scientific Research, which was the key body for the organization of scientific research in prewar Japan. Why, in that period when Japan was strengthening her military forces and entering war, was such weight attached to nuclear physics?

At the time Subcommittee 10 was formed, nuclear physics cannot be said to have had direct connection with military needs. Nuclear fission was discovered toward the end of 1938. It was not until the latter half of 1943 that Japanese nuclear physicists engaged in any atomic bomb or radio equipment project. Only in the previous year were these items included for the first time in the Extraordinary War Budget. Before this,

Table 4 Expenses of the Main Subcommittees and Special Committees of the Japan Society for the Promotion of Scientific Research, 1933–1940*

(in yen)

Subcommittee, or Special Committee	Research subject	Year of formation	Cumulative amount of expenses
Special, No. 7	Aeroplane fuel	1937	366,256
Subcomm. No. 10	Cosmic ray and atomic nucleus	1934	326,927
Subcomm. No. 1	Wireless instruments	1933	206,345
Subcomm. No. 16	National nutritional standards	1934	139,697
Subcomm. No. 19	Special steel	1934	135,945
Subcomm. No. 5	Corrosion-resistant materials and corrosion prevention	1933	130,355
Subcomm. No. 8	Tuberculosis prevention	1938	118,372
Subcomm. No. 3	Japanese encephalitis	1933	111,211
Subcomm. No. 24	Metal casting	1936	110,800
Subcomm. No. 11	Tropical life in Micronesia	1934	107,095
Special, No. 2	Economic exploitation of Manchuria, Mongolia and China	1933	101,912

* Sources: *Nihon Gakujutsu Shinkōkai jigyō hōkoku* (Projects report of the Japan Society for the Promotion of Scientific Research; 1939); *Nihon Gakujutsu Shinkōkai nenpō* (Annual report of the Japan Society for the Promotion of Scientific Research, No. 8; 1941).

nuclear physicists received no money from the military. Prewar Japanese interest in nuclear physics derived rather from its newness in the scientific world. For example, when the first cyclotron in Japan was completed in April 1937 at the Institute of Physical and Chemical Research, journalists described it as "a large and fantastic laboratory for the atomic nucleus" and took pride in the fact that it was a pioneering laboratory second only to E.O. Lawrence's in the University of California.[13]

We should not, however, hastily conclude that there was no motive for promoting nuclear physics other than a purely intellectual curiosity or a respect for pure science. Though the research sponsored by Subcommittee 10 was performed under Nishina's leadership, the initiative for forming the Subcommittee was most likely taken by Okada Takematsu. It is hardly thinkable—in prewar Japan where family-style academic cliques prevailed—that a young man like Nishina, who had no position in the Imperial Academy and held no court rank or honors, could have planned and carried out such a major project on his own initiative. In contrast to Nishina's lack of credentials, Okada, at the age of sixty in 1934, held a professorship in Tokyo Imperial University in addition to being director of the Central Meteorological Observatory and president of the Meteorological Society of Japan. Moreover, he held posts in both the Imperial Academy and the National Research Council, the two agencies that played the most important roles in science administration in prewar Japan. Okada was also often appointed to various governmental committees, such as the Board of Examiners for Air Transport. We can easily imagine how influential were his utterances on both government and the scientific community. Hence, it is reasonable to suppose that without Okada's efforts the Subcommittee could hardly have been formed, nor could such large grants for its work have been secured.

While no source materials have yet been found to clarify Okada's reasons for promoting cosmic ray research, it is possible to surmise some probable reasons. Ever since he graduated from Tokyo Imperial University in 1899 he had concentrated his efforts on modernization of meteorology and the weather services. He had long taken an interest in aerology, a pioneer branch of meteorology in those days. The importance of aerological observations and the need to grasp meteorological phenomena three-dimensionally had been stressed since about 1900, and in the 1920's reliable aerological results began appearing. Okada once said of this development: "The principles of weather forecasting underwent a fundamental transformation in this period."[14] In Japan, the Aerological Observatory was established in 1920 at Tateno, followed in the 1930's by introduction of the radiosonde, an indispensable instrument for aerological observation destined also for an important role in cosmic ray research. In 1932 Japanese meteorologists began to manufacture radiosondes. It was stressed that the introduction of physical methods was important for the advancement of aerology, and in this connection some meteorologists paid attention to cosmic rays. In a survey written in late 1932, we find the following statement:

> Since atmospheric phenomena are three-dimensional, we must take up the physics of aerology in order to gain a genuine comprehension of natural phenomena. . . . We cannot, however, satisfactorily comprehend complicated aerological phenomena without fully using modern scientific knowledge. This means, of course, thermodynamics and hydrodynamics, but also studies of sound and optical phenomena in the atmosphere, and studies of ion distribution and geomagnetism. *Recently the so-called cosmic rays have often been discussed in this connection.*"[15] [My italics.]

Before this, since about 1926, the intensity of cosmic rays had been recognized as varying with the atmospheric pressure, and a temperature dependence had also been conjectured. Connections of cosmic rays with meteorological phenomena had long been noted by many authors. The concern with aerology, then, was a natural, internal development of meteorology. But there was also a social motivation for this concern—aviation safety.

In Japan the importance of aerology to the safety of aviation had been recognized since about 1914.[16] Responsibility for aeronautical meteorology was assigned to the Central Meteorological Observatory in 1923. In 1930 the duties of the aeronautical meteorology division of the Observatory were enlarged, and in 1931 aeronautical weather charts began to be published.[17] At a time when the Manchurian Incident was about to occur and the militaristic regime grew stronger day by day, the strongest demand for development of aeronautical meteorology came, as might be expected, from the military. In May 1931 the Ministry of Education organized a committee to investigate aeronautical meteorology, the members of which were representatives of the Aeronautical Research Institute of Tokyo Imperial University, the Central Meteorological Observatory, the Army Air Force Department, the Navy Air Force Department, and the Navy's Hydrographic Department.[18]

Although we have no evidence that the growing importance of aerology and aeronautical meteorology was the conclusive reason for forming Subcommittee 10 on cosmic rays and the atomic nucleus, we may at least assume that it provided a favorable background for promoting cosmic ray research.

Nuclear physics and business circles

It was not the Japan Society for the Promotion of Scientific Research alone that financed nuclear physics in prewar Japan. The cyclotrons at the Institute of Physical and Chemical Research and at Osaka University were financed not by Subcommittee 10 but by funds from business circles.

In a 1933 report on the Nishikawa Laboratory of the Institute of Physical and Chemical Research, we find the following passage on the artificial disintegration of the atomic nucleus:

> For this experiment it is absolutely necessary to get high energy protons. Kikuchi Seishi and Nakagawa Shigeo are trying to use a transformer to get hundreds of kilovolts, but because of insufficient funds have obtained only a far lower voltage. Regrettably, various experiments cannot be carried out due to financial difficulties.[19]

As a matter of fact, in the period 1930–1933 the Institute's finances were the lowest in its prewar history. Only after Japanese industries turned to a wartime economy did they improve.[20] Experiments in nuclear physics at the Institute actually began after the construction of a cyclotron (1936–1937), for which the Institute received endowments totalling about $3,000,000 from business circles. Mitsui Hōonkai (a Mitsui-funded foundation established in 1933) offered about $150,000 for the laboratory building and accessories. The main part of the machine was contributed by Japan Wireless Telegraph Company (Nihon Musen Denshin Kaisha, established in 1925 with government finances), while running costs were covered by an endowment of about $100,000 from Tokyo Electric Light Company (Tokyo Dentō Kaisha).[21] For construction of its cyclotron, Osaka University received about $80,000 from business sources, specifically Taniguchi Industrial Promotion Foundation (Taniguchi Kōgyō Shōrei Kai).[22]

Thus, the beginning of experimental nuclear physics depended largely upon funds donated by business circles. It is noteworthy that in the mid-1930's Japanese business circles were able to make such large contributions to a field like nuclear physics, which was not then expected to yield immediately any practical results. That is, big business' funding capability rested on a new stage in Japan's economy. Seizing the Manchurian Incident as an opportunity to escape from stagnation, Japan's economy entered a boom supported by colonial exploitation and a rapidly expanding war industry. It was this boom that allowed economic circles such latitude in making substantial grants to nuclear physics.

Conclusion

Experimental nuclear physics generally incurs great expense. Prewar Japan was no exception; the amount of money spent for nuclear physics was very large compared with that spent for other branches of science. The reasons for favoring nuclear physics must be sought in the social conditions of Japan in that period.

Research in cosmic rays and nuclear physics was first sponsored by the Japan Society for the Promotion of Scientific Research and was supervised by its Subcommittee No. 10. The Society was founded in 1932 in line with the government's policy for rationalizing industry in response to the great depression, and was funded by government grants and private endowments from business circles. The Society's comparatively large allocations to research in cosmic rays and nuclear physics seems attributable partly to the novelty of nuclear

Table 5 Annual Expenses of Subcommittee No. 10*

(in yen)

Year:	1934	1935	1936	1937	1938	1939	1940
Amount:	9,136	13,335	15,150	127,850	49,929	57,726	51,800

* Sources: same as Table 4.

physics, and partly to the importance of such research to aerology in connection with military aeronautics. From the mid-1930's a booming wartime economy enabled business circles to offer large sums to experimental nuclear physics.

Abundant research funds alone do not, of course, guarantee successful research results. Due credit must be given the effective leadership of individual leaders, such as Ōkōchi Masatoshi, director of the Institute of Physical and Chemical Research, Okada Takematsu, chairman of Subcommittee No. 10 of the Japan Society for the Promotion of Scientific Research, and to Nishina Yoshio, who directed the actual research in a time in Japan when the psychological amosphere was not entirely favorable to scientific work. But, in the sense that it was, after all, the social situation that allowed them to exercise their talents, the prewar Japanese nuclear physicists were children of their time.

References

The author expresses his appreciation to Messrs. Nemoto Junkichi of the Meteorological Agency, Tsuji Tetsuo of Tōkai University, Miyazaki Yukio of the Institute of Physical and Chemical Research, and Horiuchi Gōji of the Meteorological College for their kind help in collecting source materials.

1. Sakata S., *Butsurigaku to hōhō* (Physics and method; Tokyo, 1947); and Yukawa H., Sakata S., and Taketani M., *Shinri no ba ni tachite* (In the course of our study; Tokyo 1951).
2. Yamasaki F., "Nishina Yoshio Hakushi ryakuden" (Short biography of Dr. Yoshio Nishina), in Tomonaga S. and Tamaki H., eds., *Nishina Yoshio denki to kaisō* (Nishina Yosho—his life and reminiscences; Tokyo 1952), esp. p. 8.
3. *Kagaku*, Vol. 2 (1932), pp. 527–528.
4. *Shizen*, Vol. 16, No. 2 (1961), p. 81.
5. Yukawa H., "Hansei no ki" (The first half of my life) in his *Butsurigaku ni kokorozashite* (Aspiring for physics; Kyoto, 1944).
6. Hirosige T., "Kagaku kenkyū taisei no kindaika" (Modernization of the organization of scientific research) in Hirosige T., ed., *Nihon shihonshugi to kagaku gijutsu* (Science and technology in Japanese capitalism; Tokyo, 1962).

7. Japan Society for the Promotion of Scientific Research, *Tokubetsu oyobi shō-iinkai ni yoru sōgō-kenkyū no gaiyō* (Outline of research projects under special committees and subcommittees), No. 4 (1939), p. 167.
8. Miyazaki Y., "Kaken ni okeru uchūsen kenkyū shi" (History of cosmic ray research in the Scientific Research Institute), in *Kagaku Kenkyūjo uchūsen jikkenshitsu hōkoku* (Report of the cosmic ray laboratory in the Scientific Research Institute), Vol. 1, No. 2 (1952).
9. Nishina, Y., Sekido Y., Shimamura F., and Arakawa H., *Physical review*, Vol. 57 (1940), pp. 633, 1050; Vol. 59 (1941), p. 679; *Nature*, Vol. 145 (1948), p. 703; Vol. 146 (1949), p. 95.
10. Sekido Y., *Uchūsen* (Cosmic rays; Tokyo, 1934), p. 179; Miyazaki Y., ref. 8, p. 6.
11. *Kagaku*, Vol. 20 (1950), p. 132.
12. Nishina et al, *Scientific papers of the Institute of Physical and Chemiccal Research*, Vol. 38 (1941), pp. 353, 360; Sakata S. and Tanikawa Y., *Physical review*, Vol. 57 (1940), p. 548.
13. Asahi Shinbunsha, *Asahi nenkan* (Asahi yearbook, 1938), p. 353.
14. Okada T., *Sekai kishōgaku nenpyō* (Chronology of world meteorology; Tokyo, 1956), p. 19.
15. Arakwa H., "Kishōgaku bankin no shinpō—riron kishōgaku" (Recent progress in meteorology—theoretical meteorology), in *Kagaku*, Vol. 3 (1933), p. 42.
16. Ōi S., "Nihon no kōsō kansoku no rekishi" (History of aerological observation in Japan), in *Tenki* (Weather), Vol. 2 (1955), p. 181.
17. Horiuchi G., "Okada Takematsu jiseki, IV" (Okada Takematsu's life works, IV), in *Tenki*, Vol. 4 (1957), p. 120.
18. *Kagaku*, Vol. 1 (1931), p. 136.
19. *Kagaku*, Vol. 3 (1933), pp. 486–488.
20. Ōkōchi Kinenkai (Ōkōchi Memorial Association), *Ōkōchi Masatoshi, hito to sono jigyō* (Ōkōchi Masatoshi, the man and his works; Tokyo 1954), p. 71.
21. Nagaoka H., "Sōgō kenkyū no hitsuyō" (The need for large-scale research), a speech delivered at the Japan Society for the Promotion of Scientific Research on March 31, 1937. Cf. Nagaoka H., *Genshi jidai no akebono* (Dawn of the atomic age; Tokyo, 1951), pp. 133–140.
22. Ōsaka University, *Ōsaka Daigaku niju-go nen shi* (Ōsaka University's first 25 years; Ōsaka, 1956), p. 169.

4

The Elementary Particle Theory Group*

Kaneseki Yoshinori

The "Elementary Particle Theory Group" designates a special group of Japanese scientists doing research on that subject. A summary of the group's membership and their institutional affiliations is presented in Table 1. Assembling all the materials pertinent to the group and presenting them in an orderly fashion is nearly a hopeless task. Not only do new students suddenly come into the forefront while some seniors fade into the background, but their mutual cooperation and interaction is also very complicated. Though a rough grasp of the situation is possible, constant changes in the web of historical processes are not easily comprehended. Present realities, nonetheless, demand an attempt at a precise description of the Elementary Particle Theory Group.

Judging from public reaction to Dr. Yukawa Hideki's receiving the Nobel Prize in 1949 and the flood of publicity that feeds his fame, it might be supposed that elementary particle research would have erupted like a volcano when Yukawa returned home that year, after a summer vacation in the United States. Looking back on the commotion and excitement Yukawa's award aroused among people who understood nothing whatsoever about his research, and before hailing his exploits ourselves, it is appropriate to try to discern the reasons why the prize was conferred on him. Admittedly a genius, Yukawa did not produce his theory without pain and suffering. Those who would learn something from the history of science must, then, search for the ground of Yukawa's reputation as a scientist, and this search necessarily leads to the formation and development of the Elementary Particle Theory Group.

While I had long hoped a more skillful writer would compose the history of the group, I have finally decided to do it

* Published in *Shizen* (Nature), Vol. 5, No. 4 (1950), pp. 48–55.

Table 1 Scientists Engaged in Elementary Particle Theory Research (January 1950)

Name	Professional status	Institutional affiliation	Most recently attended	Date of graduation
		University / Faculty	University / Faculty	
Shōno Naomi	Assistant	Kyushu / Science	Kyushu / Science	1947
Yamanouchi Teikichi	,,	,, ,,	Tokyo / Science	1942
Katsumori Hiroshi	Graduate research fellow	,, ,,	Kyushu / Science	1947
Fujinaga Shigeru	,,	,, ,,	,, ,,	1948
Sakuma Kiyoshi	Professor	Hiroshima College of Literature & Science	Hiroshima College of Literature & Science	1935
Ueno Yoshio	Associate professor	,,	,,	1941
Ōuchi Tadashi	Lecturer	,,	Tokyo / Science	1946
Mimura Yōichi	,,	,,	Hiroshima College of Literature & Science	1945
Suura Hiroshi	Assistant	,,	Tokyo / Science	1947
Mimura Yoshitaka	Professor	Hiroshima / Inst. for Theoretical Physics	,, ,,	1921
Kimura Toshiei	Assistant	,,	Kyoto / Science	1948
Miyachi Yoshihiko	,,	,,	,, ,,	1948
Tanigawa Yasutaka	Professor	Kobe University	Osaka / Science	1939
Fusimi (Fushimi) Koji	Professor	Osaka / Science	Tokyo / Science	1939
Utiyama (Uchiyama) Ryoyu	Associate professor	,, ,,	Osaka / Science	1940
Koba Zirō (Jirō)	,,	,, ,,	Tokyo / Science	1945

Gotō Ken'ichi	Assistant	Osaka / Science	Osaka / Science	1941
Watanabe Shigeru	”	” ”	” ”	1943
Nakajima Hiroshi	Graduate student	” ”	” ”	1946
Mugibayashi Nobumichi	”	” ”	” ”	1948
Kotani Tsuneyuki	”	” ”	” ”	1946
Watanabe Yōiti (Yōichi)	”	” ”	” ”	1947
Nakai Shinzō	Graduate research fellow	” ”	” ”	1945
Nakano Tadao	Graduate student	” ”	” ”	1948
Taniuchi Toshiya	”	” ”	” ”	1946
Suzuki Yoshio	”	” ”	” ”	1947
Nanbu Yōichirō	Associate professor	Osaka / Science Municipal	Tokyo / Science	1942
Hayakawa Satio (Sachio)	”	” ”	” ”	1945
Yamaguchi Yoshio	Lecturer	” ”	” ”	1947
Okayama Daisuke	Professor	Osaka College of Dentistry	Osaka / Science	1939
Hirano Minoru	Associate professor	Wakayama University	Nagoya / Science	1944
Shimooda Hiroichi	”	Nara Women's College	Kyoto / Science	1948
Yukawa Hideki	Professor	Kyoto / Science	Kyoto / Science	1929
Tamura Matsuhei	Associate professor	” ”	” ”	1927
Inoue Takeshi	”	” ”	” ”	1941
Hosoe Masanao	Assistant	” ”	” ”	1941

(Cont'd on next page)

(Table 1 cont'd)

Name	Professional status	Institutional affiliation University / Faculty	Most recently attended University / Faculty	Date of graduation
Tokuoka Zensuke	Assistant	Kyoto / Science	Kyoto / Science	1947
Enatsu Hiroshi	”	” ”	” ”	1944
Tanaka Hajime	”	” ”	” ”	1946
Kita Hideji	”	” ”	Tokyo / Science	1946
Munakata Yasuo	”	” ”	Kyoto / Science	1948
Hasegawa Hiroichi	Graduate research fellow	” ”	” ”	1948
Takano Yoshirō	Supernumerary researcher	” ”	” ”	1948
Hayashi Chūshirō	”	” ”	Tokyo / Science	1942
Tsuda Hiroshi	”	” ”	Kyoto / Science	1946
Hiroishi Shōhei	”	” ”	” ”	1944
Ataka Yasushi	”	” ”	” ”	1949
Kobayashi Minoru	Professor	” ”	” ”	1933
Noma Susumu	Lecturer	” ”	Tokyo College of Literature & Science	1937
Takagi Shūji	Assistant	” ”	Tokyo / Science	1945
Sawada Katsurō	”	” ”	Kyoto / Science	1947
Katayama Yasuhisa	”	” ”	” ”	1947
Shōji Seiichi	Supernumerary researcher	” ”	” ”	1947

Kawaguchi Osamu	Supernumerary researcher	Kyoto / Science	Kyoto / Science	1948
Sakihama Keiichi	Graduate research fellow	„ „	„ „	1949
Hori Shōichi	„	„ „	„ „	1949
Araki Gentarō	Professor	Kyoto / Engineering	Tokyo College of Literature & Science	1932
Sakata Shōichi	Professor	Nagoya / Science	Kyoto / Science	1933
Umezawa Hiroomi	Assistant	„ „	Nagoya / Engineering	1947
Shimazu Haruo	Graduate research fellow	„ „	„ / Science	1945
Togano Shigenori	„	„ „	„ „	1946
Kawabe Rokuo	„	„ „	„ „	1948
Oda Nobuo	Assistant	„ / Engineering	Kyushu / Science	1946
Gotō Hideo	„	Gifu / Medicine & Engineering	Nagoya Technical Institute	——
Inoue Takeshi	Associate professor (part-time)	Nagoya / Science	Kyoto / Science	1941
Hara Osamu	Assistant	„ „	Nagoya / Science	1946
Ogawa Shūzo	„	„ „	„ „	1947
Yukawa Jirō	Research assistant	„ „	„ „	1947
Yamada Eiji	Graduate research fellow	„ „	„ „	1948
Miura Susumu	Trainee	„ „	„ „	1946
Gotō Shigeo	Associate professor	Gifu / Medicine & Engineering	Kyoto / Science	1946

(Cont'd on next page)

(Table 1 cont'd)

Name	Professional status	Institutional affiliation University / Faculty	Most recently attended University / Faculty	Date of graduation
Suzuki Tan	Associate professor	Yokohama National / Arts & Science	Kyoto / Science	1943
Nishino Isao	Assistant	,, ,,	Tokyo College of Literature & Science	1948
Mutō Yoshio	Professor	,, ,,	Kyoto / Medicine	1935
Yamanouchi Takahiko	Professor	Tokyo / Science	Tokyo / Science	1926
Katō Tosio (Toshio)	Assistant	,, ,,	,, ,,	1941
Iwata Giichi	,,	,, ,,	,, ,,	1938
Fukui Masao	,,	,, ,,	,, ,,	1944
Miyamoto Yoneji	Graduate student	,, ,,	,, ,,	1945
Fukuda Hiroshi	,,	,, ,,	,, ,,	1946
Takeda Gyō	Graduate research fellow	,, ,,	,, ,,	1946
Tani Sumio	,,	,, ,,	,, ,,	1946
Kinoshita Tōichirō	,,	,, ,,	,, ,,	1947
Endō Shinji	Graduate student	,, ,,	,, ,,	1947
Tamura Tarō	,,	,, ,,	,, ,,	1949
Takayanagi Kazuo	,,	,, ,,	,, ,,	1949
Aizu Akira	Graduate research fellow	,, ,,	,, ,,	1945
Nishijima Kazuhiko	Graduate student	,, ,,	,, ,,	1949

Machida Shigeru	Graduate student	Tokyo / Science	Tokyo / Science	1949
Uchiyama Tadazumi	”	” ”	” ”	1945
Ida Kōjirō	”	” ”	” ”	1949
Okabayashi Takao	”	” ”	” ”	1949
Kotani Masao	Professor	” ”	” ”	1929
Nakamura Seitarō	Assistant	” ”	Kyoto / Science	1941
Fujimoto Yōichi	Graduate research fellow	” ”	Tokyo / Science	1947
Ono Ken’ichi	Assistant	” ”	” ”	1943
Umezawa Minoru	Graduate student	” ”	” ”	1949
Kanazawa Hideo	Associate professor	” / Liberal Arts	” ”	1941
Usui Tsunemaru	”	” ”	” ”	1945
Mutō Toshinosuke	Professor	” / Institute of Science & Engineering	” ”	1928
Inoue Kenzō	Assistant	” ”	Kyushu / Science	1944
Inoue Takeo	”	” ”	Tokyo / Science	1942
Watanabe Ichie	Research fellow	” ”	Kyushu / Science	1947
Tanifuji Makoto	”	” ”	” ”	1947
Sebe Takashi	”	” ”	” ”	1947
Tomonaga Shin’ichirō	Professor	Tokyo University of Education	Kyoto / Science	1929
Miyajima Tatsuoki	Associate professor	”	Tokyo / Science	1939
Tati (Tachi) Takao	Lecturer	”	” ”	1940

(Cont’d on next page)

(Table 1 cont'd)

Name	Professional status	Institutional affiliation	Most recently attended	Date of graduation
		University / Faculty	University / Faculty	
Sasaki Muneo	Assistant	Tokyo University of Education	Kyoto / Science	1941
Suzuki Ryōji	”	”	Tohoku / Science	1941
Kanezawa Suteo	”	”	Tokyo College of Literature & Science	1942
Baba Kazuo	”	”	”	1946
Fukuda Nobuyuki	”	”	Hokkaido / Science	1944
Ōno Mamoru	Graduate student	”	Tokyo College of Literature & Science	1948
Itō Daisuke	Assistant	”	”	1947
Hanawa Shigeo	”	”	Hokkaido / Science	1941
Mano Kōichi	Graduate student	”	Tokyo College of Literature & Science	1945
Iida Yoshikatsu	”	”	”	1948
Takahashi Yasutarō	”	”	”	1948
Watanabe Masaru	”	”	”	1948
Abe Etsuko	”	”	”	1946
Kawaguchi Hiroko	”	”	”	1944
Watanabe Satosi (Satoshi)	Professor	Rikkyō / Science	Tokyo / Science	1933
Toyoda Toshiyuki	Associate professor	” ”	” ”	1943
Fujita Fumiaki	Assistant	” ”	Osaka / Science	1947

Taketani Mituo	Professor		Kyoto / Science	1934
Nogami Mokichirō	Professor	Gakushūin University	Tokyo / Science	1936
Ozaki Masaharu	Associate Professor	Tohoku / Science	Tohoku / Science	1937
Sasaki Seibun	Graduate research fellow	„ „	„ „	1947
Oneda Sadao	Contract researcher	„ „	„ „	1946
Nakabayashi Kugao	Professor	„ „	„ „	1932
Nomoto Morikazu	Graduate student	„ „	„ „	1945
Kokubu Hidenori	„	„ „	„ „	1949
Satō Iwao	Associate professor	„ „	„ „	1944
Hasegawa Wataru	Contract researcher	„ „	„ „	1946
Sugawara Masao	Graduate research fellow	Hokkaido / Science	Kyushu / Science	1947

myself. Considerable effort has gone into this essay, though some thoughtless errors and careless mistakes may still remain. There are perhaps materials whose use would have provided a broader, more detailed picture, but I have not utilized them. Apologies may be due for not going further into the question of how to revitalize the academic world, from a historical point of view. I do expect, however, to revise this essay appropriately at some future time.

A star-studded research group

Elementary particle theory is the most fundamental level of research in the field of basic science. As shown in Table 1, more than a hundred scientists in Japan are presently engaged in research on elementary particle theory. While plausible explanations for this impressive growth of researchers in this specific area may be forthcoming, it most definitely did not result simply from uncontrolled mass production of researchers, nor from their hasty recruitment in response to—and certainly not in anticipation of—Yukawa's fame.

From the dates when particle theorists graduated from their universities, one finds that postwar graduates are more numerous than the prewar group. Because the numbers are so great, it is not possible to list all dissertations on elementary particle theory. Not a few young scientists in this field have done outstanding research soon after their graduation from university. In this connection, it is enlightening to examine the journal *Soryūshi-ron kenkyū* (Studies in elementary particle theory), published by the Japan Physical Society. This journal is quick to publish new and provocative research by previously unknown men whose work often surpasses that of well-established scientists. At the time Yukawa received the Nobel Prize, J. Robert Oppenheimer said, "Yukawa's prediction of the meson has born splendid fruit in the last ten years, and he has played an impressive role in developing high-level physical research in Japan." To what could the development of high-level research refer if not to the formation of our elementary particle group, which regularly produces important newcomers? Yukawa's achievement did not simply represent the good fortune of one individual; rather, its significance is much greater precisely because it grew out of the collective work of the Elementary Particle Theory Group.

Besides the younger men, there are also some older scientists active in the Elementary Particle Theory Group. Deeply concerned with various other matters from time to time, they nevertheless were the key members in laying the group's foundations. It has not been easy for a new basic field of physics to get into the antiquated physics departments of our universities, where something like arteriosclerosis has set in, and even now [1950] scarcely a chair, professorship or lectureship in elementary particle theory can be found. There even are some universities where obstinate authorities pointlessly try to deny it any place. As Table 1 makes clear, a good number of universities have no professor, associate professor or lecturer in this field. Research is being carried forward exclusively by very young scientists. While many senior scientists' names are mentioned, it cannot be said that they invariably are doing research in this field at the present time—or even that they did so in the past. In some cases, the senior scientists merely patronize and direct the research of younger men whenever convenient —a practice not at all unusual in Japanese universities. In fact, the role of seniors in the group has been to cultivate with care and patience the tender shoots of elementary particle theory, despite this kind of intellectual freeze. In an atmosphere marked by the unrelenting obstinacy of unthinking academic bosses, the younger scientists working today have every reason to be grateful for the determined struggle of their seniors.

The research pattern common in Japan—in which a researcher is bound to a supervisory professor by the iron-like chains of feudalistic master-apprentice bondage—has been torn asunder in the field of elementary particle theory research. By way of illustration, an examination of research reports and papers appearing in the *Soryūshi-ron kenkyū* or read at the annual conventions and executive meetings of the old Japan Physical Society would reveal how many papers have been co-authored by researchers from different institutions. Many of the research workers in this field presently based in the same institution actually came from different universities and even specialized in various fields of science. Especially is this the case with Tokyo University of Education (Tokyo Kyōiku Daigaku) and the former Tokyo College of Literature and Science (Tokyo Bunrika Daigaku), which had the characteristics of the inner sanctum of the defunct "higher normal schools." The latter university has been completely transformed into a unique, wide-open place of study. The University of Tokyo graduates comprise the majority of its younger scientists, who work under the direct guidance of Tomonaga Shin'ichirō or Miyajima Tatsuoki, or under the influence of Taketani Mituo and Nakamura Seitarō. Moreover, young research workers trained at the University of Tokyo do not all settle down in their *alma mater;* many take positions at other institutions throughout the country, and at Osaka University in particular. Why so? For one thing, the University of Tokyo has yet to cast off its long-standing sophistry and opinionated bigotry, in contrast to Osaka University where efforts have been made to revive the glorious days of Yukawa, Sakata, Taketani and Kobayashi, all pioneers in proclaiming the meson particle theory. And today, what is the situation created at Kyoto University by its own graduates, young and old? Can it still boast of its brilliant achievements, given the legal restrictions imposed on the ranks of researchers during Yukawa's absence? How is it that Nagoya University, with so short an academic tradition, is producing such able young scientists? What about Kyushu University since the resignation of Mutō Shunnosuke and Nogami Mokichirō? And Hokkaido University, where Umeda Kwai has been for so long? Or Tohoku University, Ozaki Masaharu's stronghold?

To get at each of these questions, it is first necessary to trace the main line of development of the Elementary Particle Theory Group up to the present. A brief look at Table 1 reveals that a large number of young physicists are presently crowded into positions quite unsuitable for effective research.

The golden years of the Institute of Physical and Chemical Research and of Osaka University

Let me here present a brief history, making some comparisons between the groups of scientists in Table 1 and the information presented in Figure 1. The Institute of Physical and Chemical Research was conceived shortly before World War I and became a reality on March 20, 1917 just before the war ended. H.I.H. Prince Fushimi was named president of its board of trustees, and from the scholarly community Nagaoka Hantarō, Ōkōchi Masatoshi, Ikeda Kikunae and Itō Jinkichi participated in its founding. Such eminent participation in this national enterprise reflects the social tensions existing at that time. The progress of science faithfully conformed to the impulsive demands of official policy for national prosperity and military might. The young Nishina Yoshio, who entered the Institute the following year, had already graduated from Tokyo Imperial University and remained for two years' graduate work. After three years at the Institute he was sent to Europe for advanced study. Outwardly, at least, this seems to have been an extraordinary promotion for a young engineer like Nishina naturally ambitious to rise in a world dominated by capitalism's growth. From the fall of 1921 Nishina studied physics for about one year under Lord Rutherford at the Cavendish Laboratory of Cambridge University. From the winter of 1922 to the spring of 1923 he immersed himself in theoretical physics at Göttingen. In a rather brief period of time the one-time engineering student's interest shifted to experimental physics and then to theoretical physics. Before leaving Japan he himself would not have anticipated this change.

Except for the winter of late 1927, which he spent in Hamburg, Nishina remained in Copenhagen for nearly six years, from the spring of 1923 to the autumn of 1928. These six years were a time of extraordinary development in quantum mechanics. During these prime years of his life from age 31 to age 38 Nishina worked under Niels Bohr, a master of quantum physics, around whom gathered scientists later to gain world renown. Two papers which he co-authored with I.I. Rabi and Oskar

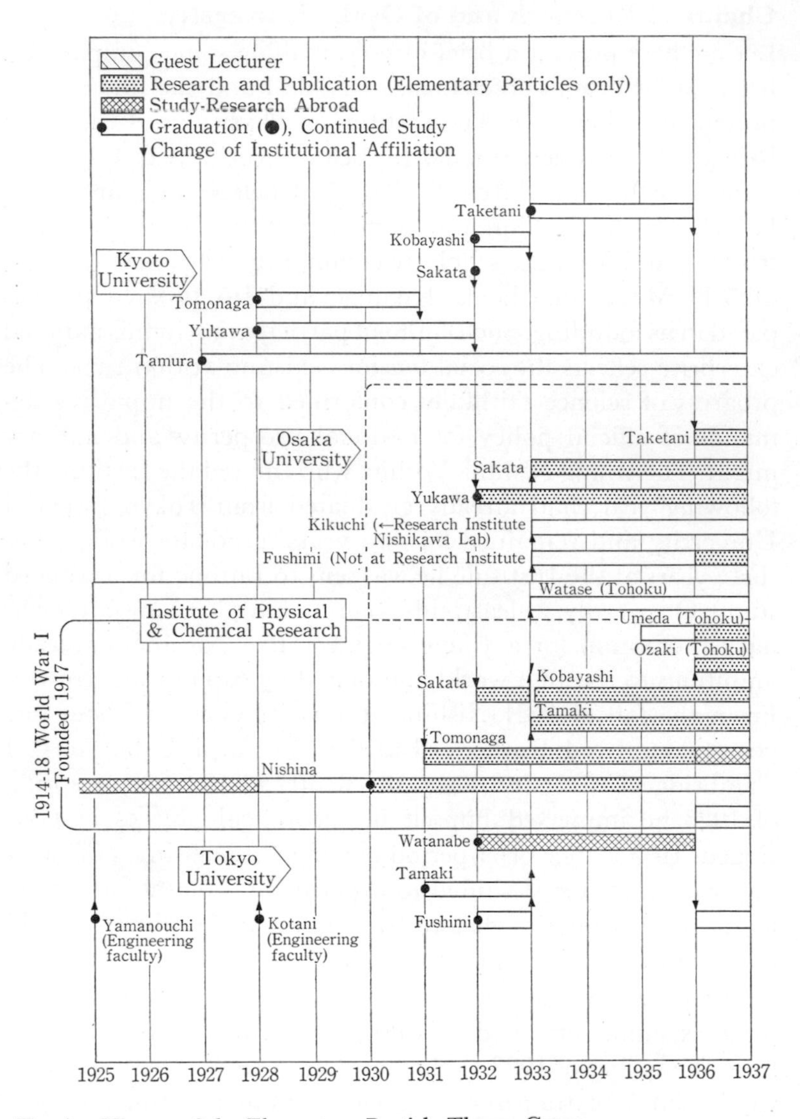

Fig. 1. History of the Elementary Particle Theory Group

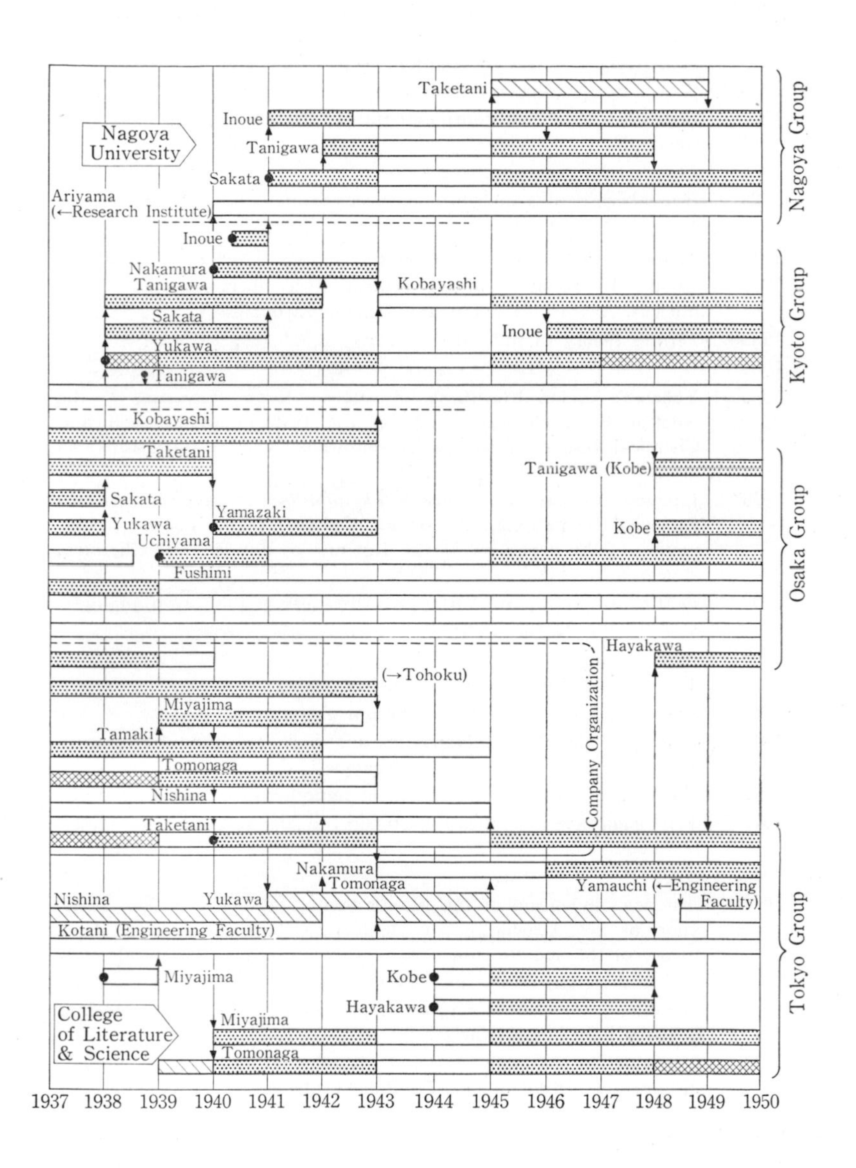

Nagoya University
Taketani
Inoue
Tanigawa
Sakata
Ariyama
(←Research Institute)
Inoue
Nakamura
Tanigawa
Kobayashi
Sakata
Inoue
Yukawa
Tanigawa
Kobayashi
Taketani
Tanigawa (Kobe)
Sakata
Yamazaki
Yukawa
Kobe
Uchiyama
Fushimi
Hayakawa
(→Tohoku)
Miyajima
Tamaki
Tomonaga
Nishina
Taketani
Company Organization
Nakamura
Tomonaga
Yamauchi (←Engineering Faculty)
Nishina
Yukawa
Kotani (Engineering Faculty)
Miyajima
Kobe
Hayakawa
College of Literature & Science
Miyajima
Tomonaga
1937 1938 1939 1940 1941 1942 1943 1944 1945 1946 1947 1948 1949 1950
Nagoya Group
Kyoto Group
Osaka Group
Tokyo Group

Table 2 History of Elementary Particle Theory (accompanies Fig. 1)

1925	Heisenberg's matrix mechanics, Dirac's quantum mechanics
1926	Schrodinger's wave mechanics
1927	Bohr's principle of complementarity, Heisenberg's principle of uncertainty
1928	Dirac's relativistic electron theory
1929	Heisenberg-Pauli's quantum electrodynamics
1930	Lawrence's cyclotron
1931	Founding of the Nishina Laboratory, Osaka University established
1932	Chadwick's discovery of the neutron, Heisenberg's model of the nucleus, Anderson's discovery of the positron
1933	Fermi's theory of nuclear *beta* decay
1934	
1935	Yukawa's meson hypothesis of nuclear forces, founding of the Nuclear Research Laboratory at the Institute of Physical and Chemical Research (cooperative venture of Nishina and Nishikawa groups)
1936	Japanese translation of Dirac's *The principles of quantum mechanics mechanics* (by Nishina, Tomonaga, Kobayashi and Tamaki), completion of the cyclotron at the Institute of Physical and Chemical Research
1937	Bohr's visit to Japan, Anderson's discovery of the meson, inauguration of the Meson Seminar
1938	
1939	The Bohr-Wheeler theory of nuclear fission
1940	Seaborg's discovery of transuranium elements
1941	Reform of the Physics Department at Tokyo College of Literature and Science, establishment of Nagoya University
1942	Sakata and Tanigawa's two-meson hypothesis, Fermi's atomic pile
1943	Tomonaga's renormalization theory, disbanding of the Meson Seminar
1944	
1945	America's invention of the atomic bomb, Japan's surrender, formation of the Association of Democratic Scientists, MacMillan's theory of the synchrotron
1946	Democratization of the Physics Department of Nagoya University, founding of the journal *Progress of theoretical physics*, Sakata group's cohesive force meson theory
1947	Confirmation by the Bristol University group that *pi* mesons disintegrate into *mu* mesons, Tomonaga's dispersion theory, Yukawa's nonlocal field theory
1948	Discovery of the meson particle by the Berkeley group
1949	Establishment of the Japan Science Council, Yukawa awarded the Nobel Prize for Physics, explosion of the first Soviet atomic bomb
1950	Announcement by the United States of the first hydrogen bomb

Klein respectively in 1928 and 1929 gave considerable stimulus to Japanese theoretical physics, which had declined somewhat since the earlier work of Nagaoka Hantarō and Ishiwara Jun. In the autumn of 1928 he traveled for a time in the United States and then returned to Japan. This was two years before Lawrence conceived the cyclotron. What Nishina would do in Japan, what would come out of his experiences abroad, could not be predicted in advance.

In 1931 Nishina became chief researcher at the Institute of Physical and Chemical Research and also founded the Nishina Laboratory there. One year earlier he had become a Doctor of Science. He engaged in introducing the latest developments in physics for three years until the spring of 1933 as a lecturer at Tokyo College of Literature and Science and for the following two years at Hokkaido University. These were probably his best years as a theorist, when he sustained his theoretical work at the highest international level, together with Tomonaga Shin'ichirō, Sakata Shōichi, Tamaki Hidehiko, Kobayashi Minoru, and Umeda Kwai. A graduate of the faculty of engineering, he remained completely indifferent to the university background of his subordinate researchers, whether Tokyo Imperial University, Kyoto Imperial University, or wherever. The spirit of Bohr, who had attracted the most promising young men of modern physics, pervaded Nishina's laboratory as well. In scale his following would not compare with that surrounding Oppenheimer at Princeton in recent years, but that does not negate the epochal role played by Nishina's group in such a provincial country as Japan. Nos. 1842 through 1845, Volume 34 (1938) of the *Scientific papers of the Institute of Physical and Chemical Research* issued as a *festschrift* for Dr. Ōkōchi Masatoshi, aptly summarize the contributions to quantum theory made by the Nishina group during these early years.

At this time Yukawa Hideki was still an unknown young scientist. His first research report, "Kaku-nai denshi no mondai ni tsuite" (On problems of electrons in the nucleus) was presented to the annual convention of the Japan Mathematico-

Physical Society held in Sendai in April 1933. He developed this paper further and in November 1934 read it to a regular meeting of the same society in Tokyo under the title "Soryūshi no sōgō sayō ni tsuite" (On the interaction of elementary particles). After February 1935 it began to circulate abroad by inclusion in the Western-language *Proceedings of the Japan Mathematico-Physical Society*. This was the world's earliest prediction of the existence of the meson particle, and Yukawa's subsequent Nobel Prize is said to have been based on this maiden paper. This sort of dramatic event is historically quite rare. On this period of Yukawa's career the reader should consult Yukawa's own *Butsurigaku ni kokorozashite* (Aspiring after physics; 1944) and *Me ni mienai mono* (Unseen things; 1945), and especially his essay "Omoidasu koto" (Reminiscences) and Sakata Shōichi's article "Yukawa riron hatten no haikei" (Background to the development of Yukawa's theory) in the March 1947 and August 1946 issues respectively of *Shizen*. Yukawa was invited to the new but highly productive Osaka University by Yagi Hidetsugu, director of the university's physical laboratory and himself a graduate of Tohoku Imperial University's electrical engineering course. From his own *alma mater* Yukawa brought Nukiyama Heiichi and Chiba Shigetarō and others to join those, including his own pupil Okabe Kinjirō, working to build up the electrical communications laboratory. As Yagi was not at all like the conventional dictatorial *n*th-generation boss of a long-established shop, Yukawa found his new working environment highly stimulating, in contrast to his more than seven years in the musty atmosphere of Kyoto Imperial University.

Successive installments of Yukawa's article "Soryūshi no sōgō sayō ni tsuite" were prepared jointly with Sakata Shōichi (second installment, 1937), Sakata and Taketani Mituo (third, 1938), and Sakata, Taketani and Kobayashi Minoru (fourth, 1938). Sakata had graduated from Kyoto Imperial University in 1933 and, after assisting Tomonaga Shin'ichirō at the Institute of Physical and Chemical Research for one year, came to work with Yukawa. Taketani had graduated from Kyoto

Imperial University in 1934, where he remained for three years before he moved to Osaka Imperial University. Kobayashi, who was about the same age as Sakata, replaced him at the Institute, but after four years left to work with the very active Yukawa group. Kobayashi and Sakata were only at the Institute for a brief time but both left papers done under the direction of Nishina and Tomonaga. Taketani had formulated his own theory of dialectical materialism with the help of progressive friends at Kyoto, and occasionally gave methodological advice on research to Yukawa and Sakata. Doing nuclear experiments under Professor Kikuchi Seichi at Osaka University in those days were Yamaguchi Tasaburō, Nakagawa Shigeo, Kumagai Hiro, Watase Yuzuru and Fushimi Kōji; their work was very closely related to the theoretical studies of the Yukawa group. Kikuchi had left the laboratory of Nishikawa Masaharu at the Institute of Physical and Chemical Research where electron diffraction photographs had been made since 1928. He had just moved to Osaka University as a young full professor in 1933. Osaka Imperial University fortunately was free from the academic barriers based on the chair system and laboratories organized around graduates from the same university or other such affiliations.

It cannot be said that the theoretical prediction of the meson immediately attracted the foreign scientific community's attention. It was very encouraging to Yukawa, therefore, to have the warm support of Nishina and Tomonaga, who readily recognized its importance. It is not surprising that a cautious man like Bohr, who visited Japan in the spring of 1937, declined to accept Yukawa's theory, as it lacked sufficient empirical evidence. Before Bohr had even returned to Copenhagen, Carl Anderson and Arthur Neddermeyer on the one hand, and Street and Stevenson on the other, working independently, found meson tracks in cloud chamber photographs of cosmic rays. Yukawa promptly published a memorandum in the *Nihon Sūgaku Butsuri Gakkai kiji* (Proceedings of the Japan Mathematico-Physical Society) to the effect that he had predicted this earlier. Oppenheimer, who had studied showers of cosmic rays, concluded that the solid element in cosmic rays was the meson predicted by Yukawa, and from this conclusion he and Serber explained the observations of Anderson and others. Stückelberg also offered a similar paper. Thereafter the meson theory gradually attracted world attention. All this shows just how bold Yukawa's prediction was. Some years later, in 1947, a group of scientists at the University of Bristol in Britain was able to show from special photographic emulsions that the meson disintegrates into *pi* and *mu* forms. Yukawa had predicted the *pi* form, while Oppenheimer predicted the *mu* form discovered by Anderson and the other experimentalists. That twelve years elapsed before Yukawa's initial prediction could be confirmed only served to heighten its significance.

Nationwide discussions on elementary particles

Twice a year in the fall and spring Kikuchi, Yukawa and others from the Kansai area (Osaka, Kyoto, Kobe and surrounding districts) would attend scientific meetings at the Institute of Physical and Chemical Research. By October 1937 the prospects for acceptance of Yukawa's theory had brightened with the discovery of the meson particle, and the Yukawa group was invited to the Institute for an open-ended discussion of the subject. This was the first of many such discussions about the meson which subsequently took place at lecture meetings sponsored by the Institute and at annual meetings of the Mathematico-Physical Society. As presentation of research findings was the main purpose of these meetings, there was never enough time for questions and answers. A free and open discussion was essential for analyzing common problems and proposing solutions. This meson discussion group was also called the "Meson-kai" (Meson Seminar). Successive meetings convinced participants of the effectiveness of such an approach and, conversely, of the unsatisfactory research conditions elsewhere. In 1935 the Nishina and Nishikawa groups collaborated in forming the Nuclear Laboratory in the Institute of Physical and Chemical Research. Scientists from both groups participated in its seminars, as did Sugiura Yoshikatsu, Tomiyama Kotarō and Mutō Shunnosuke from other research groups in the Institute, and Yamamoto Hideo from outside it.

When Dr. Tamaki Kajurō, a professor of theoretical physics at Kyoto University, died in 1932, Yukawa moved to Kyoto in the following year as his successor, along with Sakata and Tanigawa Yasutaka. Kobayashi and Taketani remained at the Institute of Physical and Chemical Research. This split imposed rather unfavorable conditions for research on both the Osaka and Kyoto parts of the Yukawa group. Yukawa spent three years at Kyoto with Tomonaga. Both had begun study of quantum theory under Tamaki and continued it after graduation. But this professor of classical physics hardly encouraged their pursuit of quantum theory. Thus, they both worked at it on their own and eventually left Kyoto, Yukawa going to Osaka University and Tomonaga to the Institute of Physical and Chemical Research. They had been preceded at Kyoto University by able men like Tamura Matsuhei, a graduate of Kyoto who remained there as a lecturer, and Nishida Sotohiko, who continued his graduate studies there, but neither had developed their talents to the degree Yukawa and Tomonaga had. The Yukawa group at Kyoto University produced Nakamura Seitarō, Sasaki Muneo and Inoue Takeshi and others, but these men usually moved elsewhere to carry on their work. Later on, Kobayashi and Taketani, as well as Fushimi, Uchiyama Ryoyu and Yamazaki Junpei, were all doing research at Osaka, but they no longer worked together as a research group. Given this unfortunate situation, Sakata Shōichi left for the newly established Nagoya University to pursue his own research goals, together with Tanigawa and Inoue. Successful democratization of the entire physics department of Nagoya University indicates the extent of Sakata's influence. Despite Kikuchi's authority and Yukawa's reputation, still one cannot quite say that Osaka and Kyoto Universities today have managed to overcome past traditions. One must, therefore, give due credit to senior scientists like Miyabe Naomi and Ariyama Kanetaka, as well as younger ones like Miyahara Shōhei and Sugawara Takashi, who lent support to Sakata's efforts at Nagoya. For details, see Sakata's article "Kenkyū to soshiki" (Research and organization) in the September 1947 issue of *Shizen*.

In 1947 Taketani Mituo left Osaka Imperial University and entered the Institute of Physical and Chemical Research where he has since displayed remarkable organizational abilities. The meson theory at first enjoyed extraordinary success, but as research on cosmic rays and the nucleus progressed, its shortcomings became apparent. To this crisis Taketani responded with a thesis in which his attention focussed on three aspects of elementary particle theory: the appropriateness of the model, limits to the application of quantum theory, and the pros and cons of various mathematical methods (see pp. 29–35, this book). Work on these problems came under the direction of Tomonaga, then in charge of theoretical research in Nishina's laboratory. In 1943 he divided the responsibilities for each of these areas, putting Tamaki and Taketani in charge of models, Araki Gentarō and Sakata in charge of applications of quantum theory and their limitations, while retaining for himself direct responsibility for mathematical methods. The research results which were reported and discussed were remarkable, even though they took place under hectic wartime circumstances. Findings were recorded in detail in Volume I of the *Soryūshi-ron kenkyū*, edited by the Elementary Particle Theory Group, and while the themes are naturally quite specialized, many of the papers still deserve attention as illuminating sources in the history of science. In fact, no philosophical criticism of Taketani's methodology of science can be done adequately without closely examining this book.

Discussions of the meson, however, eventually ceased as participants in the Meson Seminar were successively forced into senseless wartime research. Work on the meson suffered from Inoue's conscription into the army and from the impairment of Taketani's health stemming from his imprisonment. Meantime, the United States' strategic bombing almost paralyzed Japan's productive capacities and soon brought about her total defeat. The traditions of the Meson Seminar, however, were in no way impaired. No sooner had the war ended than the Seminar immediately reconstituted itself. Compared with foreign research, our level of research seemed quite high and, in fact, once reports of foreign research began to reach us, our confidence was restored.

Development of the Elementary Particle Theory Group

In 1949 Taketani demonstrated not only that the two-meson hypothesis put forward in 1942 by Tanigawa, Nakamura, Sakata and Inoue foreshadowed the observations of the Bristol physicists, but also that their formulation was superior to the 1947 theory of Hans Bethe and R. E. Marshak. Tomonaga, Miyajima Tatsuoki, Tachi Takao, Koba Jirō and Sasaki Muneo completed the renormalization theory in 1943. It preceded in time and was superior in content to the theory of Paul Dirac and Julian Schwinger. The first organizational triumph of the postwar period involved research on the cohesive power of mesons, and this began with a seminar in the fall of 1945 at the evacuation center of Nagoya University. Taketani was able to join the group's discussion at Fujimi. This research was developed by Sakata, Tanigawa, Inoue, Hara Osamu, Takagi Shuji, Hirano Minoru, Tatsuoka Makoto and Umezawa Hiroomi, and it is presently competing for recognition with the f-place theory of A. Pais. Also, Samb and Rutherford published precise measurements of the hydrogen atom's quantum level, which the Tomonaga group was able to explain admirably by freely utilizing the renormalization theory. Out of this work came the so-called Tomonaga-Schwinger theory in quantum electrodynamics. This theory was not the result exclusively of work done by the Tomonaga group; rather, they shared credit for it with American scientists working under more favorable economic conditions. The achievement on the Japanese side should be regarded as one of diligent mutual cooperation among different groups.

In terms of personnel, the Tokyo group, with its twin pillars of Tomonaga-Miyajima and Taketani-Nakamura, is definitely on a par with the Nagoya group centered on Sakata and linked to the Kyoto group through Inoue. At Tokyo University of Education, with which Miyajima and Tomonaga are affiliated, is Fujioka Yoshi, chairman of its physics department. Fujioka is eagerly awaiting Tomonaga's return to Japan from Germany to extend the scale of research in his department. From the University of Tokyo there have come more contributions to the theory of matter than to elementary particle theory. Yamanouchi Takahiko, with not a few followers, and Kotani Masao have had the important task of correcting malpractices existing there since the time of Tanaka Tsutomu. The University of Tokyo has offered guest lectureships to such first-rank scientists as Nishina Yoshio, Yukawa Hideki and Tomonaga Shin'ichirō, though they have not always been well-received there. Once when Yukawa served as part-time professor at the University of Tokyo, he expressed his willingness to receive full-time appointment. This did not materialize, however, and the circumstances ought not to be forgotten by the persons responsible. It is said loyal sons do not desert poor families, and unworthy talent is gradually leaving the University of Tokyo. The time has come for a freer system of exchanging professors and attending lectures in each university in Japan, and at the same time the University of Tokyo should take the initiative in a thoroughgoing reform of its chair system in physics. A certain layman who happened to read Sagane Ryokichi's *Genshi bakudan no hanashi* (Story of the atomic bomb; 1949) lamented that merely replacing the faculty at the University of Tokyo would not be enough—the entire university needs overhauling.

As for the postwar period at the Institute of Physical and Chemical Research, it reorganized itself two years ago, becoming the Institute of Scientific Research, Inc. Data in Table 3 relative to the Institute's atomic research during the time of its greatest productivity may elicit a strong reaction from the reader. How the prewar Institute flourished, and what will become of the postwar Institute, Inc., are questions of crucial interest for scientists concerned with the history of Japanese science and technology. Those aware of the Institute's earlier contributions to elementary particle theory may well lament the termination of its role in this area. However, certain ideas and organizational practices developed at the Institute have now spread to other Japanese institutions, and it is to them that we should direct our attention. The giants who emerged from this system included Yukawa, Tomonaga, and others; new scientific stalwarts may well emerge five or ten years hence. Yukawa and Tomonaga both grew up in families known for their intellectual achievements, and by hard work became the eminent scientists they are today. The diligence of a Yukawa, as indicated in his first paper, and the character of a Tomonaga, who educated himself as he moved from Nishina and the Institute to Leipzig and Werner Heisenberg, learning much in the process, constitute an invaluable wealth for an impoverished Japan. Younger scientists working in elementary particle theory can take heart from the patience and fortitude of their predecessors, as they will doubtless overcome difficulties of their own. The already apparent difficulties are no less onerous than earlier obstacles, but can be removed one by one by the organized efforts of the Elementary Particle Theory Group in its determination to share the agony and joy of the Japanese people that lie ahead.

Elementary particle theory is not, of course, the whole of physics. It would certainly be unfortunate if it were the only field productively active. But for ignorance and incompetence to render growth of the Elementary Particle Theory Group impossible, especially after the worldwide recognition granted Yukawa, would be a travesty of history. The responsibilities of science administration cannot be exercised without popular support, and the Japanese people are becoming alert to this.

Table 3 Nuclear Researchers at the Institute of Physical and Chemical Research in 1942

Personnel	Status at the Institute	Subsequent professional history
The Nishina Laboratory Group: Studies of Cosmic Radiation, the Nucleus and Meson Theory		
Nishina Yoshio	Research fellow	Director of the Institute for Scientific Research
Tomonaga Shin'ichirō	Research fellow	Professor at the Tokyo College of Literature & Science, professor at Tokyo University of Education
Yukawa Hideki	Research fellow	Professor at Kyoto and Tokyo Universities, termination of joint appointment
Tamaki Hidehiko	Assistant	Research fellow of the Institute for Scientific Research
Araki Gentarō	Research associate	Professor at the Tokyo College of Literature & Science, professor at Kyoto University
Umeda Kwai	Research affiliate	Professor at Hokkaido University
Sakata Shōichi	Research affiliate	Professor at Nagoya University
Kobayashi Minoru	Research affiliate	Associate professor at Osaka University, professor at Kyoto University
Ozaki Masaharu	Research affiliate	Associate professor at Tohoku University
Miyajima Tatsuoki	Research affiliate	Associate professor at Tohoku University
Watanabe Satosi (Satoshi)	Research affiliate	Associate professor at Tokyo University, professor at Rikkyo University
Tanigawa Yasuyuki	Research affiliate	Lecturer at Kyoto University, professor at Kobe University
Okayama Daisuke	Research affiliate	Professor at Osaka College of Dentistry
Taketani Mituo	Research associate	Lecturer at the Tokyo College of Literature & Science
Chon P'yong Su	Research associate	Returned to Korea
Inoue Takeshi	Research affiliate	Lecturer at Nagoya University, Associate professor at Nagoya and Kyoto Universities
Nakamura Seitarō	Research affiliate	Assistant at Tokyo University

The Nishina Laboratory Group: Quantum Theory of Solid State Physics

Ariyama Kanetaka	Research affiliate	Professor at Nagoya Uviversity

The Nishina Laboratory Group: Studies of Cosmic Rays

Nishina Yoshio	Research fellow	Director of the Institute for Scientific Research
Ishii Chihiro	Assistant	Staff member of the Northeast Scientific Institute of the People's Republic of China
Sekido Yatarō	Assistant	Professor at Nagoya University
Takeuchi Masa	Assistant	Professor at Yokohama University
Iio Makoto	Assistant	President of the Iio Electric Company
Shimamura Fukutarō	Assistant	Technical expert at the Central Meteorogical Observatory
Miyazaki Yukio	Assistant	Assistant research fellow at the Institute for Scientific Research
Inoki Masafumi	Research associate	Assistant research fellow at the Institute for Scientific Research
Kameda Tadashi	Research associate	Assistant at the Institute for Scientific Research
Miura Isao	Research associate	Assistant at the Institute for Scientific Research

The Nishina Laboratory Group: Studies of the Transmutations of Elements and Artificial Radiation, Construction of Large-Scale Cyclotrons, Neutron Research

Nishina Yoshio	Research fellow	Director of the Institute for Scientific Research
Sagane Ryōkichi	Research fellow	Professor at Tokyo University
Watanabe Sukenori	Research affiliate	Staff member of the Institute for Plant Chemistry
Shinma Keizō	Assistant	Research fellow at the Institute for Scientific Research
Yamazaki Fumio	Assistant	Research fellow at the Institute for Scientific Research
Sugimoto Asao	Assistant	Research fellow at the Institute for Scientific Research
Amaki Toshio	Assistant	Professor at the Konan Higher School

(Cont'd on next page)

(Table 3 cont'd)

Personnel	Status at the Institute	Subsequent professional history
Ichimiya Torao	Assistant	Staff member of the Electrical Research Institute
Tajima Eizō	Assistant	Assistant research fellow at the Institute for Scientific Research
Miyamoto Goro	Research affiliate	Lecturer at Tokyo University, associate professor at Tokyo University
Ezoe Hirohiko	Research associate	Assistant research fellow at the Institute for Scientific Research
Miyazaki Kiyotoshi	Research affiliate	Staff member of the Japan Electric Company Research Laboratory
The Nishina Laboratory Group: Spectroscopic Studies of Light Rays Ionized by High Speed Cyclotrons		
Nishina Yoshio	Research fellow	Director of the Institute for Scientific Research
Tajima Eizō	Assistant	Assistant research fellow at the Institute for Scientific Research
The Nishikawa Laboratory Group: Studies of Artificial Radiation		
Nishikawa Masaharu	Research fellow	Professor at Tokyo University, honorary research fellow of the Institute for Scientific Research
Nakagawa Shigeo	Research affiliate	Professor at Rikkyo University
Sumoto Inosuke	Assistant	Assistant research fellow at the Institute for Scientific Research
The Nishikawa Laboratory Group: Studies of Slow Neutrons		
Nishikawa Masaharu	Research fellow	Professor at Tokyo University, honorary research of the Institute for Scientific Research
Kimura Motohara	Assistant	Research fellow of the Institute for Scientific Research
Hatoyama Michio	Assistant	Staff member of the Electrical Research Institute

The Nishikawa Laboratory Group: Studies of Artificial Transformation of the Nucleus Using High Voltage Cockcroft Generators

Nishikawa Masaharu	Research fellow	Professor at Tokyo University, honorary research fellow of the Institute for Scientific Research
Shinohara Ken'ichi	Research fellow	Lecturer at Osaka University, professor at Kyushu University

The Nishikawa Laboratory Group: Research on the Nucleus and Elementary Particles

Kumagai Hiroo	Research affiliate	Associate professor at Tokyo University

The Nishikawa Laboratory Group: Construction of Large-Scale Cyclotrons, Studies of Nuclear Fission of Uranium, Studies of Cyclotron-produced Artificial Uranium

Yazaki Tameichi	Assistant	Professor at Yamanashi University

The Kikuchi Laboratory Group (Osaka University): Studies on Neutron Scattering, Research on Artificially-Produced Radioactive Atoms, Measurement of Magnetic Efficiency in the Nucleus, Studies on the Separation of Isotopes by the Clausius-Dirichlet Method

Kikuchi Seishi	Research fellow	Professor at Osaka University
Watase Yuzuru	Research affiliate	Associate professor at Osaka University, professor at Osaka Municipal University
Itō Junkichi	Research affiliate	Lecturer at Osaka University, professor at Osaka University
Watatsuki Tetsuo	Research affiliate	Assistant at Osaka University, associate professor at Osaka University
Takeda Eiichi	Research affiliate	Assistant at Osaka University, associate professor at the Tokyo Institute of Technology
Yamaguchi Shōtarō	Research affiliate	Assistant at Osaka University, associate professor at Osaka University
Kokubu Yujirō	Research associate	Assistant at Osaka University, lecturer at Osaka University
Sakata Makizō	Research associate	

(Cont'd on next page)

(Table 3 cont'd)

Personnel	Status at the Institute	Subsequent Professional History
Suehiro Izuo	Research associate	
The Nagaoka Laboratory Group: Studies on the Interaction of Protons, Heavy Protons and the Nucleus, Research on Neutrons		
Sugiura Yoshikatsu	Research fellow	Professor at Rikkyo University
Minakawa Osamu	Assistant	Technical expert at the Central Meteorological Observatory
The Takamine Laboratory Group: Studies on Neutron Scattering		
Fujioka Yoshio	Research fellow	Professor at the Tokyo College of Literature & Science, professor at Tokyo University of Education
The Takamine Laboratory Group: Studies on Quantum Statistics		
Tomiyama Kotarō	Research affiliate	
The Takamine Laboratory Group: Spectroscopic Studies of Light Rays Ionized by High-Speed Cyclotron		
Takamine Toshio	Research fellow	
Tanaka Yoshio	Research affiliate	Lecturer at the Tokyo College of Literature & Science, professor at Tokyo University of Education
The Ishida Laboratory Group: Theoretical Studies on the Nucleus, Research on the Quantum Theory of Fields, Quantum Theory of Solid Bodies		
Mutō Toshinosure	Research fellow	Professor at Tokyo University
The Ishida Laboratory Group: Theoretical Studies on the Structure of the Nucleus, Elementary Particle Theory		
Nogami Mokichirō	Research affiliate	Professor at Gakushuin University

Influential people in the scientific community, and in government as well, are not, we trust, so shortsighted as to let problems in our heritage go unresolved.

Studies by the Elementary Particle Theory Group's members today are published in the English-language quarterly *Progress of theoretical physics* and the Japanese journal *Soryūshi-ron kenkyū*. The former was founded primarily by Yukawa at the time when all academic journals faced economic weaknesses in reestablishing themselves after the war. It was a full-scale scientific journal aimed at rapid publication of developments in Japanese elementary particle research for distribution abroad. Its influence in raising Japanese research to international standards in a short time deserves much praise. Any intellectual who loudly advocates Japan's recovery of her international position, yet denies the spectacular achievements of this quarterly, possesses, at best, only questionable perceptivity. In the fall of 1948 *Soryūshi-ron kenkyū* edited and published manuscripts scheduled for presentation at the Elementary Particle Section of the Physical Society's annual convention. The *Progress of theoretical physics* was not at that time meeting the scientists' demands for prompt publication of research results within Japan. Reading a scientific paper at a meeting suffers time limitations; speaking, listening, mutual comprehension and discussion are all difficult in such a setting. The opportunity to publish an outline of their research and the facts it uncovered in the more informal *Soryūshi-ron kenkyū* with a sense of freedom and pleasure was not only convenient but encouraging as well.

Later on, early publication also became possible in the *Progress of theoretical physics* through the efforts of Kobayashi Minoru and Okumura Kazuo, and its reputation was quite secure. But the *Soryūshi-ron kenkyū* retained a vital function by encouraging new advances that eventually contributed to developments abroad. Correspondence between foreign scientists and Japanese researchers is often printed in this journal; this stimulates our younger scientists as did once the visits of Tomonaga and Yukawa to the United States. The journal also prints extended papers by foreign scientists in reduced form, though papers of unusual importance are reproduced in their entirety. Likewise, copies of the journal containing such papers are not hoarded but are sent out to eight different research centers around the country. Responsibility for the journal was assumed by the Japan Physical Society after Nakamura Seitarō had published the first four volumes, though his burdens have not been adequately alleviated. If even a small amount of money were invested to rationalize its operations, this problem could be easily solved. Publication of the *Progress of theoretical physics*, however, will become impossible unless something is done soon about its financial situation, Yukawa's prestige notwithstanding.

At the University of Tokyo, Taketani Mituo in particular regularly introduces newly published works to his colleagues, and physicists in the Tokyo area actively participate in his seminar regardless of their institutional ties. But the senior physicists there have ignored all this; their failure to support his efforts either materially or spiritually is without question a mean-spirited disregard of scholarly ethics.

5

Marxism and Biology in Japan*

Nakamura Teiri

About 1930, just when Marxist influence in Japan became vigorous enough to affect the theories of natural science, the Japanese proletarian movement was completely crushed under constant and violent suppression by the Japanese fascists. Subject to severe social pressures, progressive intellectuals in Japan continued to place their hopes in Marxist theory and in a socialist society like that developed in the Soviet Union, as symbols of the day when they would escape from their frustrating and stifling conditions. The oppression of the 1930's, however, inevitably caused distortions in their view of Marxism. Completely cut off from revolutionary movements, it became for them a somewhat vague world view or philosophical stance, rather than a theory of revolution.

Reflecting this distortion, and also the fact that philosophical works escaped censorship more easily than class struggle theories, it was only natural that books on Marxist philosophy (including those on cultural and scientific themes) constituted the bulk of all Marxist publications (Table 1). This not only intensified the already strong attraction to philosophy among younger intellectuals in general but, paradoxically, induced many of them to turn to Marxism.

Biologists and methodology

Most of the persons who played central roles in the post-World War II activities of either the Theoretical Biology Section of the Association of Democratic Scientists (Minshu-shugi kagakusha kyōkai, or Minka) or in the formation of the Association's separate Biology Section, were born in the period 1905–

* This paper was first published as "The pre-history (*zenshi*) of the Lysenko dispute," a part of the first chapter of *Ruisenko ronsō* (The Lysenko dispute [in Japan]; Misuzu Shobō, Tokyo, 1967).

Table 1 Marxist Works Published in Two Representative Years in the Early Showa Period (1926–)*

Field	1927		1936	
	No.	% of total	No.	% of total
Philosophy	34	14	50	45
Social science	81	34	61	55
Revolutionary theory	122	52	0	0
Total	237	100	111	100

* Figures for 1927 are from the 1928 edition of *Shuppan nenkan* (Publications yearbook), edited by Tokyo Shosekishō Kumiai (Tokyo Book Merchants Association); those for 1936 are from the 1937 edition of the same yearbook, then edited by Tokyodō.

1915. The time of their youth therefore coincided exactly with the period when the socialist movement in Japan was suppressed and Marxist philosophy gained a compensatory popularity among intellectuals. Table 2 shows the age distribution in 1935 of the main members of the Theoretical Biology Section who were Marxists or were strongly influenced by Marxism.

Table 2 Age Distribution in 1935 of the Marxist Members of the Theoretical Biology Section of Minka

Age:	15–19	20–24	25–29	30–34	35–39
Number:	3	7	4	2	1

Not surprisingly, their Marxism was the kind of distorted "philosophism" mentioned above. Nevertheless, insofar as it was Marxist philosophy, it sought some relation to practice. Progressively oriented biologists tended to accommodate their philosophy to their vocation, and thus to reduce it to a tool for biological research. From this arose their "fondness for methodology." It is clear, though, that the preferences for philosophy and methodology of Yasugi Ryūichi and other central figures in the Theoretical Biology and Biology Sections of Minka helped sustain their campaign to introduce and propagate Lysenko's theory, which after World War II gained strong support among progressive biologists as a "methodological critique" of orthodox biology from the standpoint of dialectical materialism.

The prewar period yields a vivid record of the "fondness for methodology" that already characterized progressive young biologists then—a memo written by Iino Tsugio who graduated from Tokyo Imperial University's department of zoology in 1937 and soon died at the young age of 29, and a criticism of that memo by three other members of his class, Usui Masuo, Shirakami Ken'ichi and Yoshimatsu Hironobu. Though his critics were under Marxist influence, Iino was not. (Usui and Yoshimatsu, along with Yasugi and Kusano Nobuo, maintained a reading circle on theoretical biology during the war, and to it Yasugi introduced the Lysenko theory.)

Iino's opinion was that "since it is extremely difficult to comprehend the organic relationship between a particular characteristic of a living thing and all its other characteristics," the most urgent task "in the present stage of biology" was to study each particular characteristic separately from "life, the highest integration of all characteristics of living things." His critics retorted, "facts cannot be recognized as such unless the observer has a definite philosophical framework" and "what is urgently needed at present" was the formation of a correct philosophical theory of life. This dispute is interesting as a prototype of the postwar conflict between the methodological emphasis of those sympathetic toward the Lysenko theory and the positivistic thought of those who were sceptical toward it or rejected it.

Initiatives among students of materialism

Systematic criticism of orthodox genetics from the standpoint of Marxism first came from the biologists and philosophers in the Society for the Study of Materialism (Yuibutsuron Kenkyūkai; cf. III-1). Their arguments appeared in the Society's journal *Yuibutsuron kenkyū* (Studies in materialism), first published in 1932, in such papers as "An interpretation of living things" (1932) by Hosokawa Koichi (pen name of Ishii Tomoyuki), "A critique of Mendelism" (1933) and "Genetics and materialism" (1933) by Ishihara Tatsurō, "A Darwinian task in biology" (1933) by Hashikake Akihide, and "Historical outline and future perspective of biology" (1934) by Ishii Tomoyuki. These papers were synthesized in a monograph *Seibutsugaku* (Biology; 1934) by Ishii and Ishihara, published in a series on materialism.

In the section of the book on "The present state of biology," Ishii argues that modern biology's theoretical poverty is the cause of much confusion and conflict, mentioning particularly the conflict between mechanism and vitalism. Of special relevance to the postwar controversy over Lysenko's theory is this book's criticism of the mechanistic theory in genetics:

> The mechanistic theory tends to study each phenomenon of life separately from others, as fixed, lifeless things, rather than dialectically as things which function and change in relation to one another and as a whole. The mechanist sees each characteristic of living things as separate from others, assumes a gene corresponding to each characteristic, proposes mechanistic relations between genes and characteristics, and thus formulates various theories of genetics.

In the book's chapter on evolution, this view is expanded to explain Morgan's theory of genetics:

> Living things are simply the summation of characteristics which are regarded ultimately as a specific combination of genes within a cell, and therefore changes in these characteristics are thought to result mainly from the formation of new combinations of genes through mating or qualitative changes in the genes themselves. . . . Thus, the complex process of the evolution of living things is simplified and reduced to genetic changes within the cell. This conclusion derives from a mechanistic interpretation of the relationship between genes and the characteristics of living things, ignoring the complex relationship between ontogenesis (development of the individual) and phylogenesis (racial evolution, evolution of a species). The evolutionary theory that started from Mendelism underestimates the significance of

Darwin's theory of natural selection and the problem of acquired characteristics.

Another point not to be overlooked in this book is its high evaluation of biology in the Soviet Union. Ishii states emphatically that, whereas biology ought to be developed in conjunction with production, biological research in capitalist societies has become so unorganized and unsystematic that, "on the whole, fruitful work is no longer possible." Twenty-two pages are devoted to an explanation of Soviet successes in attaining an ideal linkage between production and research. While reporting an increase in yield through mating and X-ray radiation, however, he does not mention the Michurin-Lysenko theory.

From the above it should be clear that Ishii's work plainly expresses the interest of Japanese intellectuals in Marxist theory and in the development of Russian socialism, through which they remained in touch with Marxism even after the scientific socialist movement in Japan had been thoroughly crushed. Although Ishii, Ishihara and other members of the Society for the Study of Materialism did not consciously avoid a positivist attitude, their "fondness for methodology," together with an uncritical receptivity toward Soviet theories, led to a tendency toward dogmatism. This weakness, which was to characterize the Lysenko advocates after the war, had a critic even before the war in Koizumi Makoto, who in 1932 published a paper on "Epidemic tendencies" in the magazine *Kagaku* (Science). Writing under the initials "M.K.," he took to task the Yuiken group for its dogmatism, charging, "One cannot help noticing that their writings on biology are all characterized by Marxist arguments or quotations from Marxist literature, a defect which exposes their much too uncritical attitude toward theories they fail to digest, and, in extreme cases, dependence on imported ideas rather than a hammering out of their own."

Impact of postponing the Moscow International Genetics Society Meeting

Postponement of the Moscow International Genetics Society Meeting in 1937 and the controversy over freedom of research it precipitated in the Soviet Union were briefly reported in Japan in the editorial columns of *Kagaku* and *Kagaku pen* (Science notes) from February to May, 1937. The issue was taken up not by a biologist but by a theoretical physicist and outstanding science critic, Ishihara Jun. In his paper "Science and ideological struggle," published in *Kaizō* (Reconstruction) in May 1937, he expressed first his misgivings about the Nazi rejection of Einstein's theory of relativity as a "Jewish theory" and then, pointing out that the same kind of thing was happening in the Soviet Union, lays the axe to Lysenko's criticism of orthodox genetics. "Due to an excessive emphasis on practice in Soviet science," he says, "there is an unfortunate tendency to make unfounded attacks on some theories, because a theory's validity is confused with its utility." If Lysenko had in his possession facts which prove the gene theory mistaken, he challenged, let him criticize the incorrectness of Mendelian genetics, not its failure to be of practical use.

This argument by Ishihara elicited a counterattack from Hosoi Takashi (another pen name of Ishii Tomoyuki) in the June 1937 issue of *Kagaku pen.* Despite the fact that the Soviet controversy over science and the Nazi suppression of science are no more the same than are socialism and fascism, claimed Ishii, Ishihara treated them as identical. Although intrinsically interesting, the prewar Ishihara-Ishii debate's impact on Japanese biologists was mild compared to that of the postwar Lysenko dispute. The delayed shock in Japan contrasts sharply with the instantaneous arousal of passions among biologists in Western countries, particularly Britain and the United States, which led eventually to censure of Soviet science by scientists' associations in the West.

Perhaps one reason why Japanese biologists did not immediately speak out is that the details of Lysenko's position were not yet well known. But this was also true for biologists in Western countries. It would seem, therefore, that the real reason lay in the differences between social conditions at that time in Britain and the United States on the one hand, and in Japan on the other. In the background of the Ishihara-Ishii skirmish was Ishihara's concern over the stifling controls imposed on scholarly research in Japan in those days, as mobilization for war rapidly escalated. Given the tightening of controls by the Japanese government, criticism directed at the fetters placed on research in Germany and the Soviet Union could very well be interpreted as criticism of Japanese fascism. Only an independent mind like Ishihara's dared to broach the problem.

There was little chance that the majority of Japanese scientists, mostly timeservers who willingly accommodated the interests of the ruling class, would have the heart to criticize control of research even in other countries. Geneticists in particular were only too willing to trade genetics as a 'means of improving the race' for material rewards. As this is not the place to treat this matter in detail, one example may suffice. Oguma Kan, then chairman of Hokkaido University's science department, wrote in a pamphlet on " The urgent need to establish a National Institute of Genetics": "What are Germany's invasion of Poland and Italy's attack on Albania if not the naturally bestowed 'pressure of the blood' of developing races? . . . The question that inescapably poses itself is, 'What must be done to strengthen our own race?' . . . It is precisely genetic research that is most effective for racial improvement."

Postwar "democracy" had to come before orthodox geneticists in Japan exploded in criticism of Soviet biology. Though formally the same, Ishihara's voicing of that criticism during the prewar militarist period had a social significance entirely different from the postwar outburst of orthodox geneticists.

The case of Yasugi Ryūichi

Yasugi Ryūichi's maiden paper in the field of the history and criticism of biology, "Biology in modern Russia and Darwinism," appeared in 1939, the same year that Oguma proposed a national genetics research institute, and the year World War II began. In Japan the situation grew darker; progressive movements, including intellectual activities, had been almost completely stifled. Already in December 1937 writings by leftist novelists and critics had been banned. In February of the following year the Society for the Study of Materialism disbanded. Under the circumstances, most of what Yasugi wrote on biology, ideology and society had to take the form of an introduction of Soviet literature rather than a statement of his own views. It was in such an introductory form that a scholarly treatment of Lysenko's theory of genetics first reached Japan.

Yasugi had been interested in biological methodology even before his university days. He realized that the study of the history of biology was essential if he were to engage in research on biological methodology. He also developed quite early a strong interest in the social role of the natural sciences and their place in the history of thought, an interest that led him to study the history of Russian science. Having come to the history of biology from two directions, his happy encounter with Lysenko was almost inevitable. In Yasugi's 1939 paper mentioned above, Lysenko is referred to merely as having succeeded in vernalizing wheat, but in a 1940 article on "The panorama of natural sciences in the Soviet Union" (in *Chūō kōron*, No. 7, 1940), he mentioned the controversy between Lysenko and Vavilov, as well as the background of criticism by the "mechanists" and the Deborin school, in both of which, he claimed, "theory and practice are completely divorced from one another. Both are revisionists. The fallacy of the Deborin school, however, is a reflection of the fact that Soviet domestic construction at that time did not yet rest on a solid technological base. . . . This state of affairs changed with the commencement of the first five-year plan of 1928. . . . There was a changeover in the ideology of the members of the Soviet Academy of Sciences."

Yasugi first made public his views of this theory in an introductory article on "Lysenko" that appeared in 1942 in *Kagaku shichō* (Trends in scientific thought):

First of all, it is not clear how his stand can be reconciled with at least that part of modern experimental genetics which absolutely defies negation. Secondly, he has not convincingly refuted the charge that his theory is Lamarckian. It should be clear to anyone that his theory is incomplete. However, his attempt to break away from the fixed, stereotyped view of living things so prevalent among modern biologists, and to see them in a more dynamic context, has our sympathetic support. . . . Still, as long as he is unable to substantiate his theory more systematically, it will remain unconvincing to foreign scholars despite his triumphs in theco ntroversy at home. The reason why his theory is more acceptable at home is, firstly, its success in actual agricultural application and, secondly, its compatibility with, and therefore support from the basic modes of Soviet thought. Even so, as a biological theory capable of integrating its place in Soviet intellectual history and its practical success, it is still immature.

From this it is evident that, as in the case of the Society for the Study of Materialism, the basic tenor of Yasugi's evaluation of Lysenko's theory was in large part already decided before the postwar controversy over it broke out.

Agriculture and genetics in Japan

This section's title is a phrase from Morinaga Shuntarō's paper "The orthodox and non-orthodox schools of breeding" in volume XVI of *Nōgyō oyobi engei* (Agriculture and horticulture) published in 1941. The orthodox school, according to Morinaga, follows the line of Mendel, Johannsen, Morgan and Muller, and the non-orthodox school the line of Darwin, Michurin and Lysenko. After briefly explaining Lysenko's theory, he comments, in connection with the Lysenko-Vavilov dispute: "I, too, am one of those who studies breeding on the foundation laid by Mendel and Johannsen, and am not sufficiently familiar with the Lysenko school's latest research on breeding to criticize it in depth. The reason why I have attempted here to make a summary introduction of this school is that the emergence of non-orthodox thinking has, it seems to me, something to do with the strong feeling among those engaged in breeding that the theories of the textbookish orthodox school are disconcertingly ineffective." The fact that this first introduction by an agriculturist of Lysenko's genetics into Japan was linked to the barrenness of orthodox genetics warrants considerable attention.

To test the credibility of Morinaga's testimony that the orthodox theories were "ineffective," let us examine the extent to which orthodox research on breeding, done mainly at agricultural testing stations, has contributed to breeding in Japan, by focussing our attention on rice. Of the many varieties of irrigated rice that have been used in Japan since the end of the Meiji period, a number (such as Jinriki, Aikoku, Omachi, Takenari, Kameo, Bozu, and Asahi) were selected and in time became established in Japanese agriculture through the efforts of independent farmers during the Meiji period. By about 1930, when this controversy started, such varieties were being planted on an overwhelmingly large proportion of the total irrigated acreage.

Why this was so can be seen by looking briefly at the development of the government's breeding projects. For the first decade after the Central Agricultural Experiment Station was established in 1893, it was too busy making comparative tests of already existing varieties to experiment with new ones. After Mendel's laws and Johannsen's theory of pure line became known in Japan in 1932, attempts to breed new varieties by artificial mating and pure-line selection increased in frequency, yielding 540 varieties of irrigated rice. Of these, only the Riku-u No. 132 variety was used very extensively. One reason for this meager success is the debility at that time of the orthodox school; though another is that not enough consideration was given the different ecological conditions of various provinces of Japan, with the frequent result that new varieties produced by the Central Agricultural Experiment Station brought poor results.

Apart from whether or not orthodox genetics was directly

responsible for such failures, it is understandable that agriculturists seriously concerned with agricultural production were dissatisfied with the established theories of the orthodox school, when its breeding methods met with so little success.

Introduction and testing of yarovization (vernalization)
Lysenko's first and most dependable achievement was his theory of phasic development and his work on yarovization as an application of this theory. Even before his genetics became known in Japan through Yasugi and Morinaga, his name was associated in Japan with this research. Experiments on vernalization had already been carried out since 1932 by Yamamoto Kenkichi, Tejima Torao, Kakizaki Yoichi, Kido Mitsuo, Takasugi Narimichi, Shibuya Tsunenori and others. According to Itō Yoshiaki, in the *Nōgyō oyobi engei* (1936) alone there were eight papers or reviews directly related to the subject. By about 1940, however, research on yarovization rapidly declined, in favor of research on plant hormones.

Criticism of Mendelian genetics by Tokuda Mitoshi
Shifting our attention from agriculture to fundamental biology, and particularly genetics, it is not necessary to paint a full picture of those times from the standpoint of the history of biology. Let us simply trace the circumstances that led to the defection from the orthodox school by Matsuura Hajime and Tokuda Mitoshi, two of the leaders of the Michurin-Lysenko forces in the Lysenko dispute and in the Michurinian movement after the war.

Born in 1906, Tokuda belonged more to the generation of Ishii (born in 1903) than to the generation of Lysenko supporters, including Yasugi, Iijima Mamoru, Ijiri Shōji, Usami Shoichirō, Fukumoto Nichiyo and Mafune Kazuo, all born between 1911 and 1915. Whereas Ishii became Marxist under the influence of a socialist movement on the verge of collapse, Tokuda did not receive his Marxist baptism in his student days. After graduating from Hokkaido University, he did research on the genetics of rodents and arrived at experimental results at variance with the Mendelian ratio of segregation. The American geneticist Green, engaged in the same kind of research, explained such results by means of the multiple gene theory, but Tokuda, unsatisfied with such explanations and skeptical of the "corpuscular," or gene theory, chose to leave the experimental laboratory.

In a summary of a report prepared for the Genetics Society of Japan in 1939 (but not actually presented, due to "compelling circumstances"), Tokuda wrote:

> A broad scanning of variations in animals and plants reveals that, on the whole, clearcut Mendelian segregation seldom occurs. It appears with some regularity only in plants and animals whose selection is dependent upon artificial factors. Heredity not characterized by such clearcut segregation is generally explained by means of the multiple gene theory. . . . Of late it has become clear that a variety of chromosome aberrations are involved in the formations of the species, and at last data has come forth which prompts second thoughts as to the wisdom of modern experimental genetics having completely ignored cytoplasm in its excessive emphasis on nucleoplasm. . . . Once such partial chromosome aberrations, cytoplasm aberrations and other similar phenomena occur in unison, analysis by means of the multiple gene theory becomes impossible.

Not only is this criticism of the multiple gene theory clearly a criticism of the gene theory itself, it also reveals the budding of the Lysenkoist view that it is the cell as a whole or chromosomes as a whole, and not as a collection of separate genes, that are involved in heredity. Furthermore, that Tokuda's thinking was heading toward recognizing the inheritance of acquired characteristics can be perceived in his 1941 paper on "Classification of rodents native to Japan and Manchuria." In this paper he acknowledged that mutations can cause interspecific evolution. He did not think, however, that all genetic variations were due to mutations. Rather, he also recognized geographic and directional variations as being hereditary:

> One cannot doubt that geographical variations arise from

> the direct influence of environmental conditions. But, how are such variations related to changes in the cytoplasm and nucleoplasm? At present we are unable to present any conclusive arguments in this matter. . . . Frankly, my own present thoughts are that it is virtually impossible to attribute geographic variations to the mutations so often talked about in experimental genetics, for geographic variations always correspond to environmental conditions, whereas mutations do not. (Furthermore, directional variations differ from both geographic variations and mutations.) They only become apparent over very long geological periods. . . . One must, then, recognize the fact that they are outside the scope of experimental genetics.

Matsuura Hajime's heterodoxy

Matsuura Hajime, who once served as president of the Michurin Society for a while from 1955, was Tokuda's senior, having been born in 1900. Like Tokuda, he was not influenced by Marxism in his youth. He does seem, however, to have possessed a strong critical spirit concerning biological problems in his university days. In a report to one of his teachers, Fujii Kenjirō, he ventured the thesis that, since in the pteridophytes the haploid becomes the porthallium and the diploid becomes the main body of the plant, this fact can be fully explained neither by cytoplasmic differences nor by differences in the polyploid of the chromosomes. Clearly, the spore is destined to form the porthallium, and the egg cell the main body of the plant. "This leads one to believe that the main constituent of life is hidden in the cell, with the chromosomes being a kind of shadow cast by the real thing." He revealed here his doubts that only chromosomes are responsible for heredity.

One of Matsuura's well-known achievements is his explanation of the chiasma type observed in the metaphase of meiosis of the reproductive cell, advanced in 1935 on the basis of observations of *Oenothera*. This explanation, which ran counter to the mainstream of cytogenetic thinking, made necessary a different explanation of the the mechanism of crossing over.

Furthermore, since it was this mainstream thinking about crossing over that formed the basis of Mendel-Morgan orthodoxy in genetics, it is no wonder that Matsuura became dissatisfied with the whole system of orthodox genetics. When in 1939 he discovered a variant of *Phaseolus*, the somatic segregation of which seemed not due to mutation, his doubts about orthodox theory deepened.

At that time there were some reasons for the dissatisfaction of Tokuda and Matsuura with regard to orthodox genetics. Their indication of the defects of cytogenetics, which had passed its zenith and was clearly failing to make further progress, and of the weakness of population genetics, still in its infancy, was not without significance. However, instead of following the line suggested by Tokuda and Matsuura, genetics thereafter showed vigorous growth in the direction of experimental research on the true nature of genes believed to be located in the chromosomes, and gradually overcame the weaknesses it had shown in the 1930's.

From our cursory analysis of these various factors—the spiritual ties of progressive Japanese intellectuals with the Soviet Union, the Marxist scientists' fondness for methodology, the ineffectiveness of orthodox genetics in agriculture, and the temporary stagnation of fundamental genetics—it is quite clear that the groundwork was thereby being laid for Lysenko's theory to convince the progressive Japanese scientists after the war.

Editor's Note

The original text of the *Ruisenkō ronso* is followed by five chapters in which the author discusses and critically appraises the postwar Lysenkoist and Michurinist disputes in Japan. We briefly recapitulate the major aspects as summarized in his papers "Nihon no Ruisenko ronsō" (The Lysenko dispute in Japan), *Kagakushi kenkyū*, 1964, no. 70, and "Nihon no Michurin undō" (The Michurin movement in Japan), *ibid.*, no. 71.

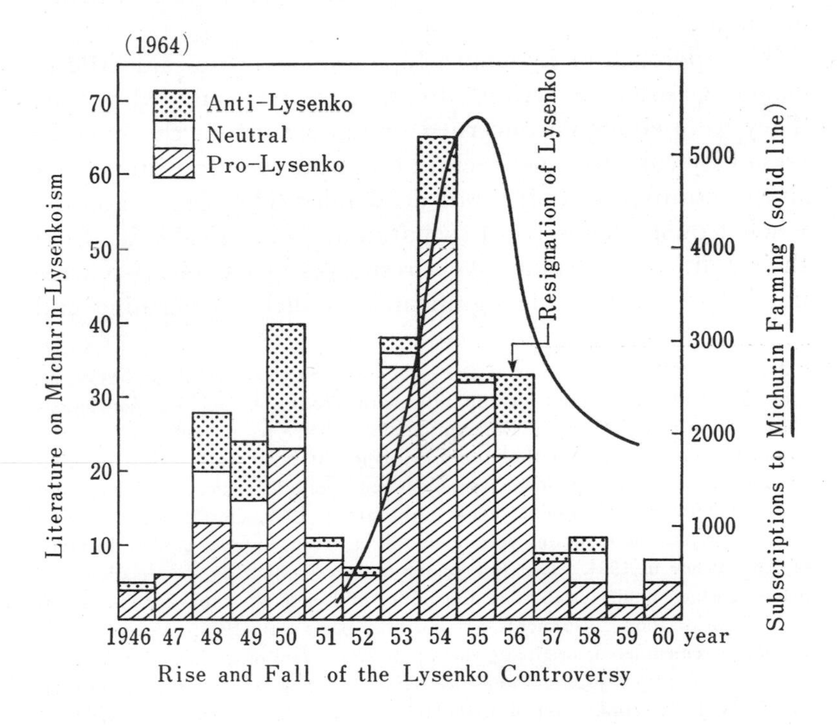

Rise and Fall of the Lysenko Controversy

There were two conspicuous peaks in publications embodying the Lysenkoist disputes (see the above diagram). In the first peak, between 1948 and 1950, progressive biologists—whom Nakamura describes above as "philosophically oriented"—faithfully digested and uncritically advocated Lysenko's doctrines, were gravely criticized by orthodox academics, and counter-attacked.* This peak declined as criticism tended to focus on the political implications of Lysenkoism. By 1951 there was little enthusiasm still in evidence. Lysenko's supporters never went beyond importing and advocating the new methodology to develop new lines of biological research based on it. Thus conventional biologists found it easy to dismiss the issues as ultimately non-scientific in character. In the beginning open-minded and uncommitted people joined the polemic with considerable excitement, but they dropped out in 1950 when the ideological significance of the Lysenko issue became apparent.

The approach of Lysenko's Japanese advocates did little to encourage sustained serious attention by open-minded people. They were either unconcerned or unable to put the hypotheses of Lysenko to experimental test, and were thus unpersuasive to positivistically inclined academics. They did not on the whole trouble themselves to separate the biological issues from their political accretions. As a result orthodox scientists were not led to consider the significance of such far-reaching and

---

* It was in 1948, at a session of the Lenin Academy of Agricultural Sciences, that viewpoints on genetics incompatible with Trofim D. Lysenko's convictions about the primacy of the environment in inheritance were prohibited in the Soviet Union. In reading the discussion below it is helpful to keep in mind that, due to Lysenko's political position, open scientific criticism of his work and opinions became possible again only late in 1952, shortly before Stalin's death. Despite the mounting volume of this criticism, Lysenko gained the favor of Khrushchev in 1954. The full scientific investigation by the Academy of Sciences which finally established the falsity of Lysenko's claims for the success of his hybridization experiments did not take place until 1965. Vernalization, it will be remembered, originally meant the treatment of grain under specified conditions of temperature and moisture before planting to alter the maturation cycle. With the application of Lysenko's approach to a much wider variety of agricultural problems, the word gradually come to be used vaguely for all sorts of pre-treatment of seed. The term "Michurinism" was used freely for Lysenkoist biology even though the views of the horticulturist I. V. Michurin (1855–1935) were by no means identical with those of Lysenko; in fact Michurin showed himself receptive to Lysenko's bugbear, Mendelian genetics, toward the end of his life. For further orientation, see Loren R. Graham, *Science and philosophy in the Soviet Union* (New York, 1972), and Zhores A. Medvedev, *The rise and fall of T. D. Lysenko* (New York, 1969). -Gen. Ed.

remarkable phenomena as vegetative hybridization. Nor were incentives to reconsider biology from the viewpoint of the interests of the people effectively communicated to geneticists.

The second peak was between 1953 and 1956. In this period the literature on the Lysenko issue was almost exclusively devoted to the encouragement of vernalization, which had reached the attention of progressive intellectuals during the first peak. The leaders of the Michurin movement recommended vernalization as an easy and inexpensive method for increasing yields that would be attractive to poor farmers. They were backed by the Japan Communist Party, whose science policy was to urge scientists to serve and cooperate with the masses.

Hindsight suggests several reasons for the failure of the movement. Agricultural yields rose rapidly in the mid-1950's, but no influence of vernalization was perceptible. Despite this general trend, the poorer farmers who were the movement's major supporters increasingly gave up trying to maintain their marginal positions, abandoned their farms, and met the demand for labor in the growing heavy manufacturing and chemical industries. Finally, in 1955 the Japan Communist Party decided no longer to commit its future to a peasant revolution. Events in the Soviet Union such as the resignation of Lysenko from the presidency of the Lenin Academy of Agricultural Sciences and the restoration of the honor of his old opponent Nikolai I. Vavilov, happened after the general decline of Lysenkoism in Japan and thus had relatively little impact there.

To sum up, the shape and placement of the second peak owed little to the content of biology; from the decline of the first peak the endurance of Lysenko's theories no longer depended upon their academic merits. The social trends that supported them were ultimately responsible for their downfall. Still it is the responsibility of progressive biologists that their enduring impact was so slight. The outcome could have been different only if they had balanced their efforts at popularization and education with scientific testing and exploration of the possibilities of Lysenko's hypotheses.

6

Grass-Roots Geology*

Ijiri Shōji and the Chidanken

Nakayama Shigeru

Of the various societies embraced by the Association of Democratic Scientists (Minshu-shugi kagakusha kyōkai, abbreviated as Minka), the only one still quite active is the Society for Corporate Research in Earth Science (Chigaku dantai kenkyūkai, or Chidanken) founded in 1947. The Historical Section of Minka, a similar organization with close relations to Chidanken, also remains in existence by issuing its journal *Rekishi hyōron* (Historical review). The reason why these two particular organizations have so far survived is found in an element common to both of them, namely, that they consist of local field researchers. This is particularly true of Chidanken, which serves as a meeting place for "grass-roots geologists," the typical member being a provincial teachers college graduate now teaching in a primary or secondary school.

There are, of course, many other organizations which promote nationwide cooperation for localized studies. Good examples are the ethnographic study group inspired by the late Yanagida Kunio and a group of amateur astronomers organized by Yamamoto Issei. In the history of science we often encounter the birth of a new research paradigm in some marginal group outside any established academic structure, but not all such groups are necessarily as radical and progressively-minded as Chidanken. Many local historical societies are marked, rather, by an unsophisticated conservatism. The reason why Chidanken still maintains a powerful progressive orientation is due largely to the personal character of its able organizer. The man who has sat in the seat of charisma for two decades since the end of World War II is Ijiri Shōji.

Chidanken is not merely a scientific society but a crusading

* Published in *Shisō no kagaku* (The science of thought), Vol. 5, No. 50 (1966), pp. 100–106.

body out to propagate its ideology and methodology. In such a group, not only charisma but also a 'bible' is indispensable equipment. To meet this need, Ijiri wrote his *Koseibutsugaku* (Paleontology, 1949), which was reprinted in 1954 under the title *Kagakuron—koseibutsugaku o chūshin to shite* (On science—centering on paleontology) and later reissued without the subtitle. Readers were immediately struck with the the distinctive individuality of the author and, at the same time, could find the cornerstone of his organization strategically placed in the treatise.

Philosophy of science based on geology as a paradigm

The major topics of the philosophy of science during the early half of the 20th century have focussed mainly on new conceptions generated from problems in modern physical science, such as causality, complementarity, and a relativistic interpretation of space-time. Ijiri's philosophy of science is, however, unique to the extent that it is heavily colored by the real experiences of a mountaineer geologist, as distinctly contrasted to armchair contemplation. Here we can note a bold stand against the superiority of exact science and mathematico-physical reductionism which have dominated the scientific world in the early part of the 20th century. In his view of science, mathematics has no reserved seat at all. First encounter reality through personal experience, next describe it accurately, and then classify it; this procedure may be called Baconian, but what is more significant is the historicism explicitly proclaimed. Ijiri's ultimate aim is to establish a unified methodology of science, taking geology as the model science and encompassing the biological, historical and social sciences. Exact science or physico-chemical reductionism is deliberately excluded from his scheme.

In the actual practice of science the present writer would not concede the excessive merit of 'philosophical' methodology. 'Given an appropriate methodology, then every solution will follow logically'—such a magic recipe with an almighty problem-solving power is hardly conceivable. In some quarters of

Japan, especially among young scientists and students right after World War II, methodology has been considered all-important. This illusory expectation may be a reflection of an inferiority complex of Japanese scientists, too anxious to catch up with the forefront of world research and to fill overnight the gap between the West and war-devastated Japan.

Some scientists are fond of ornamentation. Just as practitioners in science and technology in premodern Japan decorated their prefaces with *yin-yang* doctrines that had nothing to do with the content of their works, the generation of scientists brought up in the early decades of the present century also seem to have a common affection toward philosophy. Ijiri may not be exceptional. Among the various natural sciences, geology's prestige is low because of its low level of abstraction (measured by distance from tangible daily experience) and its lesser degree of mathematization. To those who suffer from the low prestige of their chosen subject of study, Ijiri's work, providing geology with 'philosophical' profundity, appears to have been received as a long-awaited gospel.

Moreover, unlike conventional philosophy of science, which tries to conform to the established norms of science, Ijiri's philosophy of science is so close to the daily experiences of practicing scientists that it exercises a far more positive influence. Influence varies, of course, according to the age and background of the reader. If one is an already established geologist, he may take Ijiri's philosophy rather matter-of-factly; it may arouse in him some sympathy but never influence his established course. If one is at the height of his productive research career, he may find some hints to appropriate in his own work. If one is a student or an immature novice researcher, he may receive it so dogmatically as to be in danger of falling into methodological inflexibility.

Method based on personal experience

Ijiri's philosophy of science starts with a method based on personal experience. In it we find unmistakable priority given to naked sense perception and intuition. By 'experiment' he means commitment to practice, eventually related to an ethic of social action. First, "One strikes his hammer into the earth, that is, performs some constructive action" and then "feels and experiences nature instinctively with his whole body and senses in the extreme." This is the fundamental scientific exercise. Such a methodology appeals most directly to uncomplicated mountain-climbers and amateurs rather than to men in the exact sciences who try to uncover the innermost structure of nature with highly sophisticated mathematical means. In geology, the barrier between a professional and an amateur is relatively low, compared to other scientific disciplines, perhaps because of its lesser degree of conceptual abstraction. A professional scientist tends to raise this barrier in order to authenticate his discipline. But as a professional Ijiri now tries to liquidate the barrier by laying stress on "scientifically unsophisticated, naked sense perception." Naturally, then, the frustrated energies of semiprofessionals and amateurs, annoyed by insurmountable professional barriers, can be effectively attracted and mobilized. This possibility was aptly expressed in Chidanken's slogan "penetrate into and educate the people" (*v-narodism*). I often come across groups of frustrated local amateur astronomers who say, "If only we could contribute something to the development of astronomy, however insignificant it may be, our efforts would be rewarded. Can you suggest any task that even an amateur can do?" It is much easier for local historians and grass-roots geologists to find appropriate topics to satisfy their zealous desires and to make the necessary organizational arrangements to carry them out. On such a footing has the corporate research method of Chidanken been formulated.

Emphasis upon hypothesis-venturing

Another feature of Ijiri's view of science is its emphasis upon hypothesis-venturing. In his view, 'logical' procedures for verifying a certain hypothesis or law have no significant place. The generation of a brand new idea—or hypothesis—is most important.

A general misconception prevalent among the non-scientific public is that scientists are all engaged in a search for the unknown. Even in the exact sciences, most scientists in reality undertake the search for the 'to-be-known' on the assumption that certain solid data or results are obtainable by following certain rules of inquiry. Except on rare occasions when one is possessed by an extraordinary idea, most scientists, by working along established lines, receive professional recognition and achieve personal satisfaction. In the postwar Japanese scientific community, this more conventional mentality has been predominant, and bold attempts to venture major hypotheses have been very scarce. This situation Ijiri found quite intolerable.

To a much greater extent than in other branches of natural science, the emphasis upon hypothesis-building is characteristic of geology, and especially paleontology, in which Ijiri specializes. In this field, as in archeology, an endless cycle of building and refuting hypotheses takes place without reaching any acknowledged certitude. One gets the impression that scientific progress based on accumulated knowledge, so characteristic of natural science, does not operate well here.

The stress on hypothesis-formation—or more exactly, paradigm-generation—closely parallels the spirit of dissenting from authority, according to Ijiri. Venturing hypotheses demands what he calls a "denying or dissenting spirit," quite the opposite of a conservative mentality founded on the accumulation of knowledge. He identifies this spirit with "class-consciousness itself." Perhaps he identifies the ruling class with the academic establishment of the University of Tokyo and the ruled with those dissenting against it. On the organizational level, the dissenting spirit is embodied in the anti-establishment, anti-ivory tower and anti-bureaucratic attitudes of Chidanken. In the sphere of action, Chidanken has been active along the ideological lines of Minka on the battlefield of the Japan Science Council, and thus has established a unique, democratic system for allocating its research funds. Ijiri's vigorous personality and the way his dissenting spirit stirs even idle minds into action are incarnated in the activities of Chidanken and pervade the organization.

Research-ism

Hypotheses are generated in the cerebral membranes of an individual. Therefore Ijiri clearly recognizes, despite his advocacy of corporate research, that research activity in science ultimately depends upon individual activity. Thus, he is not satisfied with the crude kind of Marxist interpretation of the the history of science that ties the development of science to that of society at large. In this respect he is more pragmatic than his Marxist colleagues, whose interpretations have nothing to do with the promotion of science, for in them he has found nothing real.

His is the attitude of an individual working scientist. Ijiri asserts, "The study of science should not be motivated by the utilitarian aim of exploitation; rather, one must maintain a mental readiness to commit double suicide with research." In this reference to the extreme sacrifice of loyal lovers we see a limitation which Ijiri and his Chidanken have placed on their general three-fold program for concentrating on research, education and action. That is, the thinking pattern which shapes the organization for concentrating on research—scientism itself—limits the scope of their engagement in scientific campaigns. They cannot become the core of a revolutionary movement by participating in community movements or by cooperating with labor organizations.

Historical science

In Ijiri's outlook on science, geology is closer to the social sciences than to the exact sciences, and in fact he places it at the core of the historical sciences. The goal of geology is description of the history of the earth, and hence an historical approach should be adopted as its research methodology.

Just as a mechanistic view of nature dominated the scientific community in the past, so it seems that at the present time physical reductionism dominates all the natural sciences.

Everything in nature, inorganic and organic, must eventually be explainable in terms of the behavior of molecules, atoms, and ultimately, elementary particles. Hence, there are no independent laws in either the geological or biological sciences.

Against this kind of reductionism, those who resist subordination to physics may raise the following criticism. Physics is a 'low-grade' science, since it deals with the simplest phenomena. Historical sciences, on the other hand, deal with the most complicated entities, and therefore the methodology of low-grade physics cannot be applicable to more complicated levels, which should not be, in any case, considered reducible to physics.

It is merely professional prejudice for physicists to claim that since physics deals with ultimate matter, it is the most fundamental of all sciences. Others may claim equally that the consideration of human nature is most essentially indispensable and fundamental, since it deals with patterns of human behavior. Or, the humanities may claim to be most important, as they deal with problems of the human soul. By the same token, the earth sciences may claim to be most fundamental, since mankind can survive only on this earth, at least for a conceivably long time to come.

Such debates actually do not constitute discussions of pure research methodology, as they derive largely from value judgments and professional prejudices about some presumed hierarchy among the various branches of science and scholarship. If a smaller element is more fundamental, then personal ethics and family affairs are more fundamental than national administration and international politics; but, in reality, the work of a prime minster of a nation, who presides over the more complex matters of national administration, is usually considered more significant, if not also more prestigious, than that of a village chief or household head responsible for smaller and simpler units.

Despite such comments and criticisms, a physicist can in reality perform his trade without paying any attention to developments in the geological sciences, while geologists and biologists must always watch out for and catch up with developments in physics. Field-oriented or phenomenological scientists may pose new problems and new objects of investigation to the physicist, but they do not provide him with any new methodology. Thus, all fields of natural science eventually become applied physics.

Disturbed by a dominant physical-reductionism of this sort, there have appeared quite a few scholars who lay claim to genuine biological and geological laws free of reductionism, in order to maintain the disciplinary independence of their fields. For instance, among Marxist scientists, Joseph Needham in his earlier writings asserted the existence of a unique approach to the levels of organization, the biologists' stronghold of independence, which the physicist and chemist can neglect. (See my "Josefu Niidamu-ron," in *Igakushi kenkyū* [Journal of the history of medicine, No. 17, 1965], or the English version, "Joseph Needham; organic philosopher," in Nakayama & Sivin, ed., *Chinese science* [M.I.T. Press, 1973], p. 25 ff.)

Though Ijiri explicitly claims that the methodology of the geologists belongs to the historical sciences, the method as defined by him is nothing but concentration on field work. Each historical event has its own individual characteristics in space and time. Hence, unlike timeless universal physical occurrences, historical events cannot be dealt with at one single central research institute, but must be pursued through the accumulation of field investigations at every provincial level and in each locality.

Field work can be carried out with the simplest equipment —a hammer, a clinometer, and a pair of tough feet. As far as a particular locality or particular field is concerned, those whose labors involve endless walking and exhaustive observations, whether amateurs or semi-professional geologists, know more and know it better than an established top-notch geologist settled in Tokyo. In this respect a local historian or a local researcher is able to maintain his identity and pride. Likewise, the corporate research method, by organizing semiprofessionals and students, has great merit, inasmuch as the exploration of mountains and fields requires as many human hands as possible. Thus, the 'historicism' and 'field-concentration' of Chidanken have been brought into a close relationship at its organizational level; and conversely, in order to maintain its organization intact, Chidanken must commit itself to 'historicism' as its official methodology.

Diametrically opposite the extreme of 'field-concentrationism' stands physico-chemical reductionism. The latter claims that, rather than exhaustive investigation by the time-honored 'hammer and tough feet' method, the most recent findings of experimental physics and chemistry, such as X-ray analysis and high-pressure techniques to reproduce earth history processes, should be widely adopted. In order to carry on such a high-technology approach, it is necessary to acquire sophisticated apparatus so costly as to be beyond the reach of any individual researcher's personal budget. To meet the high costs of such an approach, some sort of centralization is inevitable. A high-pressure laboratory can be installed only at one or a very few central research institutes funded by the government. Geological study at this level turns out to be unattainable for our grass-roots geologist. This issue of high-cost apparatus presents him with a problem far more serious than those faced by his fellow grass-roots historians, anthropologists and sociologists. While research in physical sciences and technology rapidly progresses toward centralization into 'big science' using high-cost research apparatus, the anti-establishment orientation of Chidanken can be understood as a reactionary stand against 'modernization'; it is forced to adhere to its 'historicism' because of the inaccessibility of costly instrumentation.

I fully recognize the significance of research pursued along pluralistic lines. At the same time, we must admit that many precariously popular modernistic approaches can easily sink into oblivion in due time. Meanwhile, the general trend toward reductionism into a physico-chemical approach invades irresistibly into the biological and other sciences alike. Biology and earth sciences can no longer remain aloof from this general trend. However much one emphasizes that geology is one of the historical sciences, the question remains: what is the specific merit of being a historical science? Therein is nothing particularly new, nor any sign of leading toward a new horizon.

It can be legitimately argued that geological science, as well as organic evolution, deals with unrepeatable historical experience, and that it therefore can be neither reproduced nor verified; that is quite unlike physics and chemistry. We have no rigorous proof that Newtonian laws operated in geological ages in exactly the same manner as now. We may also argue that the laws proper to geology are not mechanistic but rather phenomenological and kinematic.

However, as Miyashiro Akio proposes in his "Chikyū kagaku no rekishi to genjō" (History and present status of earth sciences; in *Shizen*, May 1966), we cannot prove that geological phenomena are beyond the reach of the laws of physics. In the fields of history and biology, annoying factors like free will and organism come into play, but geology is much closer to pure physical or material science, the only difference being its involvement in the kind of time sequence that historical sciences deal with. It may not be too profitable for the geologist to draw his methodological stimulus from the established historical disciplines and thus intentionally reject the intrusion of physico-chemical methods. Unless it is liberated from the limitations of the framework of historical science and is open to enrichment by whatever is useful, Ijiri's philosophy of science will itself turn out to be a historically outmoded fossil.

If one wishes to learn from historical methodology, examination of the historical development of geological explanations and hypotheses may be suggestive. Ijiri's 'history', however, remains natural history. Is it not rather odd that Ijiri and his Chidanken have not displayed as much interest in examining the history of geological sciences as have physicists in the history of physics?

The various elements discussed above may be summed up as follows:

	Philosophy of science	*Philosophy of organization*
1.	Methods based on personal experience; emphasis on practice and action	Emphasis on the role of amateurs
2.	Emphasis on hypothesis venturing	Value of the dissenting spirit and anti-establishment orientation
3.	Research concentration	Scientism
4.	Historical science	Localism, emphasis on field work, and corporate research method.

Ijiri's philosophy of science corresponds beautifully with the organizational principles of Chidanken. For this reason, it appears that both Ijiri's leadership role and the organization of Chidanken manifest a stability unparalleled in similar organizations. Ijiri's unique philosophy of science is primarily the outcome of his own personal experience; his version of dialectical materialism seems to be more or less a philosophical ornament to buttress his own thought. (See his *Chishitsugaku no konpon mondai* [Fundamental problems in geology]; Tokyo, 1949). He was not disturbed so much as other professional leftist thinkers by the criticism of Stalinism; he claims that, even after the incident, we may be able to learn something from Stalin on a case-by-case basis. His self-confidence as a competent natural scientist and as the moving force of Chidanken provides him with a solid ground undisturbed by shifting ideological vicissitudes. After two decades of its existence, though, his organization faces a variety of problems for the future.

Trouble spots in Chidanken

Chidanken embodies two different functional roles, one as a crusading movement, and one as a scientific society. Were its function limited to that of crusading, then it would have faded away along with the decline of Minka. The reason for its surprising longevity may be found in its two-fold nature: crusading activity in time of emergency, and normal research activity during less critical times.

As a scientific society, Chidanken faces the problem of confrontation with physico-chemical reductionism. It can continue to produce at least some meaningful research along normal lines, such as increasing the precision of geological maps, by retaining its present priority on field work. Such work has as much worth as the endless excavation of local historical sources. Yet, however progressive the corporate research method may be, its instrumentation remains old-fashioned and, consequently, at some time in the future Chidanken may no longer sustain its morale as a scientific vanguard.

Its parochialistic approach also has dangerous elements that may lead to narrow isolationism. At the time when the establishment of the University of Tokyo's Institute of Nuclear Physics became a controversial issue, Chidanken proclaimed its slogan of "science for the (Japanese) people" and stood in opposition to those physicists who were afraid of remaining at an internationally inferior position in the front-line competition of nuclear research.

Ijiri emphasized hypothesis-building and also at one time supported the Michurinian theory. There may be some similarity between geological hypotheses and the Michurinian one; in an area where experimental proof is hardly attainable, hypothesis-venturing is more appealing than verification. The abstraction of a hypothesis is generated out of one's pre-disciplinary life, or ideology, as Ijiri put it. In this area, there are some discernible dangers of indulging in irrationalism.

On the other hand, in spite of the emphasis upon hypothesis-building, corporate research operates in such a way as to curtail individualistic claims to credit. A highly imaginative hypothesis generated in the mind of a single individual may, in the process of mass discussion, become diluted and have its sharp edge moderated into mediocrity.

Chidanken's achievements as a crusade body have been spectacular and dramatic. In the early phase of the postwar democratization movement, it heralded the democratization of the scientific community by making some solid achievements in that direction. As a brake against possible reactionary moves in the future, Chidanken has reason enough for its existence.

Still, democratization of the scientific community does not necessarily require a Marxist methodology; it is more likely a 'bourgeois' revolution. Here we find one limitation of a scientists' movement. Once depived of its morale, it may degenerate into a mere professional society or a pressure group like a Medical Association.

In pursuing the goal of its 'science for the people' movement, the activity of Chidanken has been remarkable. However, scientists have, after all, little chance of becoming the main force of social revolution and, if isolated from the general trend of leftist movements, can hardly achieve anything significant themselves.

Furthermore, it seems that in its scientism, Chidanken imposes a certain limitation on itself in its approach to the people and society. Research-concentration or scientism may be necessary prerequistes for the professional activity of scientists, but are hardly appropriate as goals for popular movements. Would it not be possible for Chidanken to join in tackling such problems as pollution and community development in close cooperation with public health engineers and civil engineers in local government?

Chidanken advocates the dissemination of science. Then, what is to be disseminated? In an earlier age when superstition overshadowed society and no public education was available, enlightenment by science as a rationalistic form of thought was still meaningful. But, in the present age when moon rockets are launched for greater political glory, science is excessively deified among laymen. Thus, Chidanken, the last stronghold of postwar scientists' movements, now appears to have lost sight of its original goal, and its once attractive slogans sound jaded.

Generally speaking, a crusading body should not remain in existence just to perpetuate its own authority; it had better be dissolved. It was only natural that Minka and associated bodies should lose their original glory, but, then, what new vision is emerging to replace the old? With its policy of excluding members over forty years of age from its council, Chidanken metabolizes its executive officers; but how can its policies and activities be metabolized?

It may also be natural that any crusaders' group eventually becomes fossilized; and yet, there may be no compelling reason to declare its dissolution. But the next generation should proceed cautiously by analyzing and examining their inheritance, in order to avoid the failings experienced by their predecessors.

Perhaps the scientists' movements are now passing through a winter day, or worse, the hottest time of summer, when everyone is short of breath. Fortunately, Chidanken has so far survived in this suffocating climate. However, a crusading body should, by its very nature, be metabolized all the time; mere survival is not to be simply congratulated. The grass-roots Chidanken must, then, scrutinize the merits and demerits of of the Ijiri legacy, with rejuvenation for the future in mind.

Appendix

A BRIEF HISTORY OF MINKA

Because Minka played a crucial role in the scientists' movement just after World War II, we append an account of its activity mainly based on Hirosige Tetu's *Sengo Nihon no kagaku undō* (Scientists' movements in postwar Japan; Tokyo 1960).

The inaugural assembly of Minka was held in Tokyo on January 12, 1946. Most of its organizers had been active in Puroka (Proletarian Science Institute) or Yuiken (Society for the Study of Materialism) before the war. These earlier Marxist groups had virtually no contact with practicing scientists. Minka tried to organize a wide variety of intellectuals, liberals as well as Marxists, not only social scientists but natural scientists, technologists, and medical specialists. Among the natural scientists who made up a third of its membership were Sakata Shōichi, Taketani Mituo, Tomonaga Shin'ichirō, and Yukawa Hideki.

Minka attained this diversity because of enthusiasm for democracy, which was just emerging from long prewar and wartime suppression. Most of the scientists also felt responsible for their wartime cooperation with the military government, both in their research and in their public conduct. Before the acceleration of the Cold War, the General Headquarters of the Occupation was still friendly to activities which encompassed a wide band of the political spectrum. W. W. Hicks, a staff member of the C.I.E., sent a personal message of congratulations to the inaugural assembly.

At that meeting seven goals were adopted:

1. to enhance the morale of scientists and to establish democratic science;

2. to encourage, focus, and satisfy popular desire for science;

3. to oppose anti-democratic policies, ideas, and educational institutions;

4. to aid the advancement of young scientists;

5. to mobilize science and technology for the popular welfare;

6. to defend and raise the professional status of scientists and engineers;

7. to attain unrestricted freedom of scientific activity.

Thus Minka served three sorts of function: as an activist group propagating political ideals, as a professional association committed to raising the socioeconomic status of its members, and as a learned society furthering scientific research. In this Minka was similar to any modern scientific society, but the diversity of world-views and particular aims was greater even than that of the American Association for the Advancement of Science today. Through most of Minka's short life the political function was predominant. It was best known for preaching democracy, criticizing the prewar imperial system with its oligarchy and lunatic militarism, and denouncing opportunistic wartime demagogues. Its intent to unite all scientists, Marxist and non-Marxist alike, did not keep Marxist points of view from becoming predominant in its political discourse. Even in the field of natural science the favorite subjects of discussion were the historical relations of science and capitalism, criticism of prewar Japanese science, and such Marxist philosophic topics as the dialectical view of nature.

In the beginning, Minka's members were scientists living in Tokyo and its environs. Local chapters were gradually organized in major cities and then in rural areas. When Minka's membership peaked at the end of 1949, it had 110 local branches. Seven sections were included in its founding schema: philosophy, politics and economics, natural sciences, history, art, education, and agriculture. A very broad definition of "science" was clearly intended. The Natural Science Section was divided into several subsections (geology, biology, physics, chemistry, and engineering) and special groups (Quantum Theory Study Group, Theoretical Biology Study Group, History of Science Colloquium, etc.). Membership was at first limited to specialists inside and outside the universities, but later Minka broadened its recruiting to include students, and in some areas labor and the general public. Its peak membership amounted to eleven thousand.

The degree of research orientation varied from one section to another; geology and biology were strong in this respect. Minka differed most from present-day scientific organizations in its lack of activity to further its founding goal of improving its members' status and living conditions. Even though starvation was everywhere in postwar Japan, Minka made no serious effort to become a professional union. This possibility was perhaps ruled out by a preference for ideological analysis and propaganda activities. Until 1947 the greatest concern was focussed on criticizing the past. So much of Minka's attention was devoted to prewar absolutism and feudalistic land ownership that the new constitution and postwar land reform were inadequately recognized. The primacy of ideology is understandable, because the loss of the war had completely overturned the old system of values and new ones were urgently sought, but as time went on the incompatibility of propaganda and research aims led to increasingly perceptible weaknesses.

In 1950 the Korean War began, and the ascendancy of the political right in Japan was furthered by the "Red Purge," analogous in many ways to the MacCarthy purges in the United States but actively sponsored by the Occupation authorities. Many democratic associations born after the war ceased to exist. Minka survived without much damage, and in fact reached the high point of its activity. This was partly because it alone provided psychological and, within limits, effective political support for scientists threatened by budget and personnel cuts of research institutions, by the purge, and by effects of the war. Furthermore, Minka shifted its emphasis from the political to the research function, and thus broadened its attractiveness to scientists and students.

As deterioration of hope for the postwar period was deepened by the progress of the Korean War, the exclusion of Soviet Russia and China from the peace treaty, and the Red Purge, the Cadres' Bureau of Minka advocated the primacy of political action. Some members opposed this policy, which conflicted with the research functions that had attracted them to what was fundamentally a scientific organization. Eventually Chidanken took the lead and the voice which carried most authority became that of its founder, Ijiri Shōji. The influential voices advocated a threefold function for Minka: research and innovation, dissemination of scientific knowledge, and establishment of conditions for the development of science.

An urgent need was felt to create a style of scientific work which would serve the people and learn from them. In 1951 the Japan Communist Party resolved that, since the San Francisco peace treaty had resulted in greater dependence of Japan upon the United States, the struggle for independence of the Japanese people must be enhanced. Accordingly Ishimoda Tadashi of the History Section of Minka asserted that the most important task that science and the scientists' movement could take up toward this end was to create a "science for the [Japanese] people." The programs of Ijiri and Ishimoda combined to determine the populist course of Minka in the years that followed. In 1952 an editorial in *Kagaku bunka news* (Science and culture news) explained "science for the people" in terms of identifying urgent themes in the contemporary struggle of the Japanese people, taking up topics pertinent to popular needs, and promoting science in such a way as to liberate the Japanese masses. Scientists were urged to move among the people, enlighten them, and learn with them. Popular education was to lead to a new generation of workers able to realize a new science and a new culture. Although unclear concepts were no rarer here than in the "public enlightenment" programs of scientific societies today, the "science for the people" movement found a ready response among fieldworkers in history and geology. In 1952 there was a campaign among students to return to their native villages. There was considerable education and propaganda work among rural people, and even some sociological investigation.

Among natural scientists there were many different responses, depending upon specialties and many other factors.

Some recalled the notorious Nazi *Volkswissenschaft*, and others were led to potentially constructive viewpoints: that in Japan basic science is out of touch with technology, that scientists in Japan have the habit of depending upon other countries for their research topics, that copying vitiates science in Japan, and so on. But on the whole unreflective activism and the primacy of politics over science prevailed among advocates of "science for the people," and they demonstrated little ability to respond to visible popular needs.

This confusion of aims led to a noticeable decline in Minka's activity, but "science for the people" was again stimulated in 1954 by the incident of the Bikini hydrogen bomb test. In Tokyo the Physics, Chemistry, Fishery, and Geology subsections worked together to publish popular accounts of hydrogen-bomb fallout. Public esteem for scientists reached new heights at this time, and Minka scientists, although they were not the only people active, played a major role in molding better-informed public opinion.

Nevertheless by 1955 the decline of Minka had become apparent. As the Korean War prompted the relaxation of anti-monopolistic laws and led to the recovery of the giants of industrial capital, and as the research activity of professional scientists returned to normal, the call of Minka and of the Japan Communist Party for democracy lost its poignancy. As we have seen, even though the faction which supported "science for the people" was dominant in Minka, this idea never reached consensus. Fewer and fewer technical people continued to see it as urgent or realistic when prosperity removed the blocks to their personal advancement. Moreover, in 1955 the Japan Communist Party greatly modified its guidelines, moving from a dogmatic and coercive line to a flexible one. Minka's cadres, caught in the attendant confusion, had to modify their previous attitudes through "self-criticism." Although the organization never closed down, it continued to fade and by the summer of 1957 it hardly existed.

In conclusion, let us sum up its accomplishments. It aroused social and political concern not only among social scientists but among natural scientists as well. It served as a motive force in the reform of the academic professions, challenged an antiquated system of university appointments, and made the management of learned societies somewhat more open. Finally, it provided a forum and meeting-place for scientists whose interests transcended their disciplines.

7

A Basic Theory of Kōgai*

Ui Jun

For about ten years now, every time I have visited areas afflicted by *kōgai*** and talked to the victims, I have been confronted with their questions: Why had *kōgai* come? Why had it been allowed to become so severe? Why had nothing been done to prevent it? These questions frequently reflected public distrust toward our modern technology, and feelings close to hatred toward this professor coming down from the University of Tokyo to tell them, "*Kōgai* isn't so bad. Try to endure it." Placed in the position of representing my university I many times felt my responsibility for these things, and could only lower my head before the victims.

My profession concerns the treatment of waste water from factories and the development of techniques for the prevention environmental damage (*kōgai*). However, my experience as a faculty assistant in civil engineering has entirely convinced me that the only purpose of these techniques is to gain time and to silence protest. As an instructor I have been in charge of some seminars and laboratory classes at the University of Tokyo, but the only thing the students were looking for was their engineering diplomas, or the accumulation of units to receive those diplomas. The students were given my laboratory instruction because they needed units in the "testing of water quality."

---

* Appeared in *AMPO*, a report from the Japanese New Left, Nos. 9–10 (1972), pp. 15–26. This is a summary of a lecture and answers to questions that followed.
** The Japanese word *kōgai*, which is used to identify the pollution problem, literally means "public hazard" or "public nuisance." The standard translation is "environmental pollution," but the Japanese usage is much broader and refers, in addition to ordinary environmental pollution, to many other problems that are not strictly pollution : factory noise, excessive vibration, obstruction of sunlight, traffic congestion, water shortage, and so on. Here, as in the article in *AMPO* from which this essay was adapted, it is sometimes tranlated "pollution" And sometimes the Japanese word is used.

What I have come to realize through several years in this position is that at present the University of Tokyo is nothing more than a collection of vocational training centers. The law department is a vocational training center producing bureaucrats; and the engineering department, teaching applied chemistry, mining engineering, electrical engineering, shipbuilding, civil engineering—we can easily imagine the enterprise that fits each of these. Therefore, it is natural that problems of environment pollution (*kōgai*) are excluded.

When students learn science or science-based technology in this way, that is, solely as a way of advancing their own personal interests, they perceive nature as merely an object to be utilized, to be cut up into pieces for use. Attempts to deal with *kōgai* problems are limited to the treatment of waste in the case of water pollution and to the building of high chimneys in the case of air pollution. These are precisely the techniques for utilizing a nature that has been split into pieces, and thus such techniques represent only a partial response to the complex social phenomena of *kōgai*, and a response which is anyway too late. Furthermore, in my investigations, beginning with the Minamata problem, (environmental mercury poisoning), I have always found that universities and their graduates, and more specifically, engineering departments and their engineers, are always on the side of those causing *kōgai*. The University of Tokyo is the most typical example: the first and last time this university has ever done any research from the standpoint of the victims of *kōgai* was in connections with the late 19th-century mine poisoning incident at Ashio.

It is natural that a university which is given money and buildings by the government, and whose purpose is to train officials and engineers who can serve as useful personnel for the government, should teach techniques whose only use can be the expansion of the strength and wealth of that country. It is a fact that we ourselves have chosen the University of Tokyo as a place to learn these techniques. When I visited Warsaw University in Poland, I found that its classrooms were in private houses in an area similar to the Ginza. I was told that the reason for this was that the universities had been the bases of resistance every time the country had been invaded, and that they had therefore been destroyed by the occupying forces, professors commonly being shot or purged from the country. But eager students still visited the homes of their professors, where they heard lectures secretly late at night. These private homes later became what is now the university. When I heard of this, I realized the possibility of the existence of a university which is precisely the opposite of the University of Tokyo.

Now, if we exclude the element of privilege, and the element of rising in the world through university education, what will we have left? With regard to the *kōgai* problem, is it possible to create an academic discipline or a body of knowledge which does not consist of techniques for producing *kōgai* or for the establishment of countermeasures by the producers of *kōgai*, but is structured rather around a common understanding, or a common experience, with the victims of *kōgai* in Japan, how they have resisted, why they failed or how they won, and how they have progressed or retrogressed during the last 100 years? This kind of knowledge would perhaps aid no one in rising in the world, but it is knowledge that is necessary to our survival, as *kōgai* gets progressively worse.

By chance there was a room in the civil engineering department that was not in used at a certain time at night. I thought that it might not be meaningless to start an "open school" which offers no knowledge of use to the establishment in the very place where, in the daytime, establishment engineering is taught under the name of "city planning" and "sanitary engineering" and so on. So I settled on the idea of using that room, and thanks to various people we have been given permission to use it.

Those who experienced the "campus disorder" of 1969 may know that the buildings of the University of Tokyo are not the buildings of the people. They are under the control of the school authorities, and to get permission to be here we had to accept many conditions. One of the conditions was not to fight violently in this room. [Laughter] It is a joke to us, but not to the professors. So, I told them that my first promise was that I would request the members of the open school not to exchange blows with sticks.

(Answering a question): In your question, you pointed out that I began our school without first giving a definition of the term "*kōgai*," thus departing from the orthodox practice of most universities.

Well, let me tell you some of the definitions which are offered by people who are called "*kōgai* scholars." In general they are of two kinds. On the one hand there are highly technical definitions, like the definition of air pollution given by the World Health Organization. On the other hand there is the kind of definition which is offered by government or industrial agencies, or by the professors associated with them, in which it is always said that the assailant* in a *kōgai* problem is an unspecified many, and that the victim is also unspecifiable. But the expression "unspecified many" is made up of two different terms, and if we investigate adequately, we *can* specify both the assailant and the victim, even though they may be many.

So, I decided not to depend on any definition, and to take any case in which the inhabitants of an area claim, "We are victims, and we are victims of *kōgai*," as a case of *kōgai*. For ten years now I have had no time to make a definition of *kōgai*, being forced simply to follow its living and moving instances. Permit me, then, to proceed without answering your difficult question—though it is fundamental—of what definition I would propose.

Perhaps you know that Japan has the worst problem of environmental pollution of any country in the world. It is surrounded by the sea, with the tide coming in and going out,

* The term "assailant" comes from a pair of Japanese terms (*kagaisha*/*higaisha*) which mean, depending on the context, oppressor/oppressed, assailant/victim, or aggressor/victim. Professor Ui's sense of the term is a politicized one that analyzes the problem of pollution in such a way as to allow a struggle to be carried out.

and the Black Current flowing by; it is a country with much rainfall, which does not remain long on the land but flows swiftly out to sea. The biggest lake in Japan is Biwa-ko, where the water completely changes approximately every seven years. Seven years is not a very long time. In Sweden and Finland, for example, where there are lakes filled by the water from glaciers, it sometimes takes scores of years for the water to change completely. Of course, in Japan there is Mashu-ko, a lake with no exit, where the water stays a very long time. But that is an exception; in general, water flows through Japan very rapidly. If we have a lake that requires several decades for the water to change, and we pollute it, the consequence is that the polluted water will not go out during one generation of man. And, even if the water goes out, the mud and fish do not. It is well known that once a lake is polluted there is little possibility for recovery. However, in Japan if we drain waste into the rivers, the rivers rush it out to sea within ten days. And, of course, in practice that is what has been done and is being done.

The same thing can be said about the wind. There is no season, no month, no ten-day period, when there is no wind. Especially the stronger winds, which blow almost every day, wash out the polluted air, though the air pollution in the Kanto Plains (around Tokyo) never leaves completely. So, this is a country constantly cleansed by natural forces.

Nevertheless, we are living in the world's most polluted country. It is difficult to express the severity of this pollution quantitatively; but it was in Japan that various new diseases fatal to man, and coming from water pollution, first appeared. Minamata disease appeared with high incidence in two places in Japan in rapid succession, and it is reported that there has been a third, less severe outbreak in Finland. It is almost certain that the Itai- itai disease (*itai* means "pain"), which results from cadmium poisoning, first appeared at Toyama, and then later at Tsushima. The Yubimagari disease (*yubimagari* means "twisted fingers," one of the effects of this disease), which is suspected of being a kind of poisoning by cadmium and other heavy metals, has appeared in a number of places in this country. Now the Itai-itai disease has appeared in a third place, also in Japan. So, my assertion that Japan is the country with the most advanced *kōgai* has become so well accepted that even [then] Prime Minister Sato Eisaku does not deny it.

The first factor in this situation has been, I think, the rapidly developing economy. It is often said that *kōgai* is a side effect, or distortion, of rapid economic development. I think that this is a notion that comes from the side of the *kōgai* producers. As far as I can see, *kōgai* is *not* so trifling a thing as a mere side effect or distortion of rapid economic development. To say that it is a "distortion" is to say that if economic development were carried on rightly, or managed to follow a natural course without any distortion, *kōgai* would not appear. But the fact is that *kōgai* is one of the most powerful factors *inherent in* rapidly developing economies. Japanese economists have pointed out a number of factors in the success of Japan's capitalist economy, and the things most often stressed are low wages and trade protection. Now I am adding to these a third: *the neglect of kōgai*—permitting the economy to dirty its own clothes. The problem of *kōgai* is an essential part of the structure of Japan's capitalist economy.

For example, there are six or seven major pulp producing countries in the world: first, the U.S.A.; second, Canada; and followed by Sweden, Finland, the U.S.S.R. and Japan. If we compare these six countries, it is obvious that all of them except Japan have rich supplies of wood and water. Japan is a country with little wood and water, yet with a population density relative to its inhabitable land of from twenty to a hundred times that of the other countries. Nonetheless, there are few pulp factories in Japan that make any reasonable effort to treat their waste. The most blatant example is Fuji city, where the big pulp factories have been located right in the middle of the crowded city, and not one of them is making any attempt to treat its waste. There has been no objection raised, and so it has been possible to locate the factories there and develop the "rapidly expanding economy."

In other countries it has been a matter of common sense that between 10% and 20% of the production facilities are necessarily directed to treatment of waste, so that it does not do severe harm to the fish. In Japan even this 10%–20% is used for production. Moreover, factories are allowed to locate near consumer cities, or even in the middle of them, so that products can be transported at little cost. Only a fool could fail to make pulp producing into a booming business under these conditions. So, Japan is ranked among the world's first six pulp producers, where it has no business being.

Someone might object by noting that the pulp industry is not a principal industry. Then let us talk about an industry which anyone would agree is central: the steel industry.

Steel engineers say that the most important factor in the rapid recovery and modernization of Japan's postwar steel industry has been the oxygen LD furnace. We are familar with the multicolored smoke that rises from the chimneys of steel mills, which is a necessary product of the LD furnace. In Europe or the U.S.A. this kind of furnace cannot be used unless its smoke is cleaned in a dust collector. If this dust collector is taken into account, however, the LD furnace is not particularly cheap. The dust collector and the necessary connecting ducts are several times larger than the furnace itself. Of course, we cannot judge their cost merely from their size, but the cost of the collector runs to at least 30% of the whole system, and the small furnace itself is already very expensive. In Japan it has been possible thus far to operate this furnace without the cooling duct and dust collector. The Japanese steel companies have, therefore, profited some 30% on each furnace, a profit that can be turned to more profit, and so on.

The same kind of thing can be said of most other industries, large or small. This is my claim, and if our economists find that I have misunderstood the matter, they should correct me, giving concrete examples. But so far there have been no straight answers forthcoming.

The second factor is the close interlocking between government and the corporations. They help each other to make *kōgai* more severe. This close cooperation is the main theme in our open school. In the Ashio case (1880's–1890's) the government went all-out to suppress, smash, and corrupt the protest movement. For example, the police called in each activist, threatened him, and hinted at the sweet life he would live if only he cooperated with them. We can see also in the Kawamata incident that the police had no trouble smashing the protest, just in the same way they are presently suppressing the student protestors. Given this close cooperation, there is no point in trying to distinguish between the government and the assailant corporations, or between the Ministry of Industry and the Ministry of Welfare.

When the group of petitioners representing the victims of Minamata disease came down from Niigata to Tokyo, the section chief or whoever it was that received them in the Industrial Water Section of the Ministry of Industry pointed out the window and said, "In that five-storied building over there many doctors from the University of Tokyo are doing research and they say that your claim about the waste from the factory is groundless." Now, so far as I know there is absolutely no research institute or organization that has a five- or six- or seven- or eight-storied building anywhere near the Ministry of Trade and Industry. Maybe he was pointing to the Ministry of Finance, or some other government building. How quick to deceive are our bureaucrats—these bureaucrats produced by the University of Tokyo.

Another important factor responsible for the spread of *kōgai* is our poor sense of community. As Miyamoto Ken'ichi pointed out in his *Shakai shihon ron* (Theory of social capital), the number of self-governing bodies in Japan has been reduced, through consolidation, to one-twenty-seventh of what it was 100 years ago. There have been two great drives to consolidate self-governing bodies: one occurred in the Meiji period for the purpose of spreading elementary education, and the other came in the postwar period, partly for the purpose of establishing and maintaining lower elementary schools. The big aim has been to diffuse education. Standardizing the language and teaching basic calculating skills have been very important in the foundation of modern industry, by creating an efficient labor force and military manpower reserve. There is no other country in which rationalization has reduced the number of self-governing units to less than three-tenths of what it once was. Through this rationalization we have obviously lost our power to make decisions concerning our local communities; they can no longer even be called self-governing. In fact, there have been cases where local units have been consolidated specifically for the purpose of silencing the voices of protest aginst *kōgai*. This has happened again and again from the Ashio case right up to the Fuji case.

The problem of *kōgai* also involves a problem of social discrimination. We can point out various ways in which *kōgai* creates social segregation. One is, for example, afflicted by a pollution-caused disease, becomes poor because of it, and then is discriminated against because of his poverty. The reason why one group of Minamata patients accepted the humiliating compromise plan for compensation was that it was the only way to escape this discrimination. Until then they had been in the position of depending upon the "sympathy" payments from the corporation, and were treated by the townspeople as money-grubbers. Now at least the compensation is official, even though it is not eough. In Tsushima, an agreement was made among pollution victims to keep away scholars who came from the outside to investigate, because if it became known that there were victims of Itai-itai disease, no members of their families could marry or get a job.

For the victim of discrimination it is difficult to explain just where he is discriminated against and just why he feels uncomfortable. Discrimination is something pertaining to his whole life, and if he picks out just one or two phenomena, like his lower wages or his longer working hours, these will be refuted by some strong evidence. There will be some statistics produced to show equal opportunity in employment or marriage. Quantitative expression always deals only with small parts of the problem, thus breaking down the *gestalt* of discrimination against whole lives.

It is announced that sulphuric acid gas is present at such and such ppm (parts per million), and that lead is present at a level of so many ppm in our daily suffering of urban *kōgai*. We breathe air with various poisons—sulphuric acid gas, carbon monoxide, lead, or oxidants; and we drink polluted water and eat foods polluted with chemicals, or mercurous chemicals and chlorous agricultural chemicals; and we suffer from noises and vibrations; so it is not surprising that we get sick. Under such circumstances it is very hard to show a connection between a certain pollutant and, say, asthma or bronchitis. Nevertheless, in spite of this difficulty, some correlation between sulphuric acid gas and bronchitis can be found. This does not mean that, since sulphuric acid gas is present at only half the level of the environmental standard, it should not be blamed as a cause of bronochitis. The sufferer always bears *kōgai* with his whole body, that is, synthetically. The assailant or a third party judges from data which represents only a small part of the whole. Thus, the assailant's understanding of *kōgai* is at a completely different level from that of the victim. It is impossible to put these two levels of understanding on a single scale and then to take the point halfway between them as an objective or neutral evaluation. There are many scholars who arrive at their evaluations in just this impossible way. They listen to the assailant and to the victim, and take some "satisfactory" position in between. In doing this they actually take the side of the assailant or the discriminator. The only possible answer to the assailant's question about how the victims feel is to ask him to come and live with the victims, to breathe the same air, drink the same water, and thus feel the same feeling.

There have been many cases in which the University of Tokyo has taken sides with the polluters, sought to obscure the causal relations, or has even suppressed honest researchers. Perhaps the people who did these things were caught up in the system of the University of Tokyo, and subjectively did not intend to commit such acts. Perhaps this university by its nature is structured in such a way as to lead those who start with good intentions into the position of the assailant. The actual shape of research and education at the University of Tokyo is that of petty status-seeking carried on within the limited specialties with narrowly limited perspectives, and therefore rendered systematically incapable of grasping a problem such as *kōgai*, which has to do with the synthetic effect of complex and overlapping natural conditions. Thus, the scholars of the University of Tokyo often, no, *necessarily* stand on the side of the assailant.

Let me share with you an incident that occurred at the Nissan Chemicals plant in Toyama prefecture, as an example of how we engineers split nature into pieces. Nissan Chemicals was draining its waste into the Idagawa river, where it sometimes killed the fish. When angry fishermen protested against the factory, the response was that the pH meter had not moved since the evening before, and therefore the factory was not responsible. The fishermen then made two separate fish preserves, one above and one below the place where the waste entered the stream. On the following day the fish kept above that point were alive, and those below it were dead. The fishermen rushed to the factory with this finding, only to be told by the company engineer and a prefectural official that the pH meter had been operating all night, that the waste was always regulated so as to be neither acid nor alkaline, and that their pH meter was far more scientific than the fishermen's simple test with the fish.

Of course, this kind of thing goes on all the time. Perhaps you readily accept that medical drugs are more effective in various combinations, but we forget that mixed poisons are also more effective. This kind of geometrical effect is not considered in the setting of our "environmental standards." This shows how naively we believe in today's science.

An example of a somewhat advanced form of this kind of thinking was once pointed out by Miyawaki Akira. He had a discussion with a leading economist about city planning, and was given an easy-going answer: "If you need green, it is enough to paint some plastic green. It is a needless luxury to leave green trees standing at a time when land prices are so high." He insisted that if there is some special psychological effect of greenery, the color green should be enough. Today's ideal of city planning is a city which looks nice and where one can move freely anywhere by car. Perhaps the model for this will be "University City" at Tsukuba, for it is being planned by the city planners of the University of Tokyo—though it has already been decided that the University of Tokyo itself will not move there.

You are already aware that capitalism today is profiting even from *kōgai*, that *kōgai*-preventive industries are now the stars of the stock market. Some students of, for example, electronics will say that they have no connection at all with *kōgai*. However, according to some computer manufacturers, some of the best customers for middle-range computers are *kōgai*-monitoring centers. Most of the divisions of the engineering department of the University of Tokyo—and thus many engineers in Japan—have something to do with *kōgai*, and most of them are making money from it.

When monitoring by computers, or the setting of "environmental standards" is discussed, the question always comes up as to why it is that we never touch the source that *produces kōgai*. If we did, we could *eliminate kōgai* before we *measured* it. But present preventive techniques cannot do that. Our way of attacking the problem is organized in such a way as to leave the source alone, and to consume as much money as possible in "preventive measures." The procedure that has become orthodox, growing out of the technological viewpoint of the University of Tokyo, where all of us make our living, is to monitor *kōgai* with computers. But the present technological system itself is necessarily producing *kōgai*, and cannot be separated from it.

Needless to say, most engineers within the corporations are in equally degraded positions, and are treated entirely as no more than parts of a machine. I am involved presently in the Minamata disease trial in Niigata, and in court my task is to examine and refute each piece of evidence offered by the defendant corporation, Showa Denko. What I have found out in this process is that their pieces of evidence are beautifully and tightly separated according to their narrowly channeled professions. In chemistry, for example, they are separated into the specialties of analysis, medical chemistry, statistics, and chemical responses. Each specialist makes his own testimony independently of the others. Each testimony is often very plausible, that is, it persuades us that Minamata disease did not come from factory waste. But, if we listen to some three or four of their testimonies, through one full day, we can recognize the lack of consistency among them. Day after day we have been addressed by contradictory testimonies.

If the top man of Showa Denko were to come into the courtroom and hear the testimonies of his men and see how they are refuted, and if he were a man of common sense, he would immediately stop fighting the claim in court. However, in Japanese legal procedures it is permissible for the defendant never to appear in court, leaving the whole thing in the hands of his lawyer. The engineers who are appointed to testify will not report back to the company that their arguments have been refuted. In many cases they do not even realize that their testimony has been refuted, for within their narrow perspectives they feel some consistency. I suppose that the trial will go this way up to the final judgment—a time that is almost endless for the victims.*

Many believe that there exist techniques for the prevention of *kōgai*, but in fact there are no such techniques. The only

* On June 21, 1973 Showa Denko finally signed an agreement on compensation to the Niigata Minamata disease victims, agreeing to *all* the monetary demands of the victims and apologizing for not having learned a lesson from the Minamata disease case in Kumamoto in Kyushu prefecture. This came eight years after the Niigata case of Minamata disease was detected. The company also promised to negotiate with sincerity if new problems arise, and to suspend factory operations when danger is clearly indicated. Meantime, it promises to prevent recurrence of the disease.

thing taught at universities is that the most economical way to dispose of waste is to let it go out to the sea. Only when this becomes impossible, they say, should one look for some other treatment. The notion that anything diluted to its invisible state thereby ceases to exist was introduced from the U.S.A., where it has been found that, at least for the time being, no visible harm seems to result from the practice of diluting waste in the sea. For us the sea is a vital source of food, while in America fishing is primarily a kind of sport, though of course some inconveniences are experienced there from polluted water. But the American situation is not so pressing as is the Japanese, for we have almost nothing to eat if we lose the fish. Nevertheless, we have adopted the American theory that disposal into the sea is harmless. We see here how Japan's capitalist industry is willing to buy anything foreign or Western, no matter how stupid it is.

In fact, this principle, or rather, hypothesis, of dilution has been refuted in practice many times, and we have been warned that we could not depend on it. The oldest refutation was the phenomenon of the *midorigaki* (green oyster), which has been known in Europe since the 19th century. In a copper or zinc contaminated sea, oysters grow green and puckery, and this has been common knowledge since before World War II. In this case we find almost 100 ppm of copper and zinc in the oysters, but little in the water. Between the water and the oysters there takes place a sharp rise in concentration of some 10,000 times. Shellfish generally have a function of concentrating these metals from the water, a fact know since the prewar period.

The next clear refutation of the hypothesis of dilution was the radioactive tuna that appeared following the nuclear test at Bikini. It was not such a simple matter as the tuna simply being covered with the "ashes of death" and therefore becoming radioactive. The artificial radioactive material was first eaten by small plankton, these plankton were then eaten by small fish, which were eaten by larger fish, which in turn were eaten by the tuna. Through this process an extremely high level of radioactivity accumulated in the tuna, despite statements by our own goverment and by some American scientists to the effect that because the amount of the sea water was so very large, one or two bomb tests on that scale were by no means dangerous. Moreover, the accumulation of the radioactivity released at that time continues today, not only in the sea, but on land as well, especially in the Arctic Circle.

Now we are beginning to regret that we did not realize earlier, with the concrete evidence of *midorigaki* and radioactive tuna, that diluting dangerous substances in the water is not a safety measure but, on the contrary, a positive danger.

In the Minamata case we find a similar accumulation. There the density of mercury in the water is at most one microgram per liter, a level undetectable by normal analysis. But we find some 10 ppm of mercury in the fish, which is a rise in concentration on the order of 100,000 times, and sometimes as much as 1,000,000 times. Perhaps you all know that at the beginning of the 1960's natural concentration of DDT in animals was also found. But all these facts have not dissuaded our universities from teaching the theory of dilution.

Let us shift our approach somewhat. The same conclusion would follow even if we did not know about Minamata disease or about the concentration of DDT. For example, the shell of a shellfish is composed almost entirely of calcium carbonate, $CaCO_3$, and this calcium ion is gathered by the shellfish from the water, where it is present in the amount of, at most, one ppm. Several percent of our bones is calcium phosphate, $Ca_3(PO_4)_2$; the same is true of the bones of fish, but the water in which the fish live contains only 0.01 ppm of calcium phosphate. These things can be understood by common sense, without any training in chemical analysis. But present education makes this common-sense understanding altogether difficult. We engineers will look at the shells, note that they contain much calcium carbonate, and conclude that if shells are cheap maybe it would be profitable to convert them into fertilizer. It is very difficult for such a mentality to grasp that these shells were made by shellfish through an elaborate process of gathering calcium from the water. The conclusion that occurs to this pitiful state of mind is that *kōgai* at half the level of our "environmental standard" can never cause diseases.

Don't trust our present science ! Don't trust our present education ! I say this because I think that if we take this stance, have this common understanding, we will have some hope of getting out of this present situation.

(Answering a question on how standards for quality controls on waste discharges are determined): We can simply say that all the standards are nonsense; any water quality standard set according to the water quality maintenance law is nonsense. The way these standards are fixed is by settling on a level at which a given corporation can carry out treatment of its waste discharge without going bankrupt. Once the corporation presents a level and says that any higher standard would force it into bankruptcy, then the standard will be set at about that level. For example, the standard for the Tagonoura area of Suruga bay is a maximum of 70 ppm for floating substances. This level can be obtained by simple negotiation; though not even this method has been practiced before setting this standard. As another example of this nonsense, I hear that a standard has been set at a level of 1,200 or 1,300 ppm for a brewing company that had been putting out several thousand ppm up to that time.

To meet the present water quality standards, actually no treatment at all is needed. Since no limit is imposed on the amount of water that may be used, factories can simply take in a lot of clean water to dilute their waste and that suffices. There are practically no water quality standards that set a limit to dilution, except perhaps at Ashio. So, if the waste is twice as dirty as the standard permits, all that need be done is to drill some wells to get water for dilution. This method is actually practiced more than any other, and in the worst cases the waste is diluted at a rate of more than ten to one.

This is the nature of all the quality standards that the government has issued. To say that we should adopt these standards as soon as possible is the same as saying that we should immediately give official approval to pollution. Yet this is what has been demanded by practically all the political parties and workers organizations. The logical result of this way of thinking is the institution of a new law defining *kōgai* crimes in such a way as to protect the big corporations like Showa Denko or Chisso from all punishment. Such a measure would have at most only a slight effect on smaller corporations, and would most likely be interpreted to attack minor assailants like chimneys of public bath houses. In short, such a law would benefit only large capitalists, not the people. In fact, such limitations always deal with only one or two particular substances; the standards thus permit companies to drain off all other substances without limit, thereby raising the general level of pollution.

As Taketani Mituo shows in our joint work *Anzensei no kangaekata* (Viewpoints on safety), almost all of the quantities permitted by these standards are not quantities which meet a scientific standard of safety, but those which strike a social balance between drawbacks and advantages. Because these quantity standards are always imposed on those who do not know this, *kōgai* is, as I said before, a kind of social discrimination. Its full extent cannot be caught through quantification, that is certain. It is useless to talk about ppm's and we should abandon this approach entirely.

Several years ago when I was researching Minamata diseases, I came to realize that there are four stages in any *kōgai* dispute. First, pollution or damage from pollution is discovered and research into the cause is begun. Second, the cause is found. Third, the assailant company or some self-styled "third party" invariably raises objections to this finding. The fourth stage involves neutralization in which the question of which side is right becomes lost in vagueness. This process of neutralization by obfuscation is what I call the first principle of *kōgai*.

In the Kumamoto Minamata case, the first stage fell between 1956 and 1959, the second was the summer of 1959 when organic mercury was found to be the cause, and the third stage of objecting to this finding ran from that autumn to 1960. In the third stage, the emphasis is put less on the quality than on the sheer quantity of objections. Since the only purpose is to deceive people who know nothing, a wide variety of different claims can help obscure the real cause. Of course, if all the claims are obviously deceptive the goal of effective neutralization is not well served, so there are sometimes objections that appear rather scientific—for a while.

In 1959 Chisso, the assailant company, raised four different objections with gradually increasing detail. At bottom they were merely repetitions of the same claim, namely, that there was no relationship between Minamata disease and the mercury in its waste. In fact, in the previous year, when Kumamoto University scientists were still sticking to their idea that manganese or selenium was the prime suspect, Chisso put together the possible refutations of the scientists' hypothesis and announced its objections. In 1959 the same refutations were repeated with the words "manganese" and "selenium" replaced by "mercury."

Of course, there were other obviously transparent objections besides these. The most typical was that given by the managing director of the Japan Industrial Chemical Society, Ōshima Takeji. He argued that the cause might be some bombs that had been disposed of at the end of the war, a line initiated by the city mayor (who had previously been one of the chief engineers at Chisso). But this objection survived scarcely a month, because a former officer who had handled the bomb disposal testified that they had been disposed of not in the inner sea of Yatsushiro but far out in the deep sea off Sasebo. It does not matter so much if the true nature of a objection is revealed, though, as its only function is to gain time and turn aside our attention and energies.

The next objection raised was the famous one made by Kiyoura Raisaku, a professor of the Tokyo Institute of Technology, who tried to make the world believe that some ammoniate from stale fish had caused the disease. Another part of his objection was that there were some other places where fish contained much mercury but caused no disease, and that therefore mercury had no relation with Minamata disease or with the discharge by Chisso. He indicated in his report that such places really existed, but said that their names would be concealed, in order not to upset the public. Some of the places could be easily identified, but one that was supposed to be in the Hokuriku (northwestern) district could not. This report became very important for us, and it took me two years to locate the place. At last we found that it was Naoetsu, where along the upper part of a river there are two factories using mercury. The fact that the fish at Naoetsu are polluted with mercury became positive evidence in support of the hypothesis that the fish of Minamata accumulated mercury from the waste of Chisso. Kiyoura's desire to conceal the place names was quite understandable.

In 1960 the so-called Tamiya Committee was formed, having as its members many famous authorities* and claiming to be impartial. After some detailed study it turned out that this committee was financed by the Japan Industrial Chemical Society.

Meantime, at Kumamoto University, findings were reported one after the other that indicated large amounts of mercury in the fish and water in Minamata gulf. In November 1959 a special research committee of the Welfare Ministry, consisting mostly of Kumamoto University staff, issued an interim report to the Welfare Minister that the cause of the disease was a kind of organic mercury found in the patients, in experimental animals, and in the mud taken from the bay. This committee was dissolved that very same day. The dissolution was so abrupt that one of its members did not find out about it until several days later, when he heard about it from newsmen.

The following year the government formed a new committee of scholars and officials to study Minamata disease. This com-

* Ui's long account of committees investigating the Minamata case, listing names and organizations, is omitted here.

mittee met four times a year, and then dissolved, saying its budget was exhausted. The committee's reports were all kept secret.

Thus, the fourth stage in any *kōgai* struggle is characterized by joint efforts of the government and the scholars to nullify any findings inconvenient to the assailant company. This is what I call the seond principle of *kōgai*: there can be no such thing as a "third party."

Finally, in the summer of 1959 the people of Minamata learned from the newspapers that the cause of the disease had been verified to be Chisso's waste, which is what they had suspected all along. In August they held a series of demonstrations, sit-ins and bargaining meetings with the factory officials. The company still maintained its arrogant attitude. Claiming that its staff had also conducted investigations and had found no relation between mercury and the disease and that, therefore, it could make no compensation; since, however, they had heard of the poor catch of fish, they would present some money to the patients in the form of a gift. Finally, the fishermen shut up the company representatives in a room and smoked them all night with burning red peppers. The police moved in and arrested the fishermen for illegal detention. Some important persons in the city then stepped in to "mediate" between the the victims and the company. They were the city mayor and members of the municipal assembly, and they managed to force a reduction in the demand of the Minamata Fishermen's Organization from several hundred million yen (one hundred million yen at the time was about $277,777) to 35 million yen.

Compensation money is always fixed between the figures proposed by the two parties, not halfway between the two, but at the approximate geometrical average.* This is what I call the third principle of *kōgai*. If one side insists that there

* The geometrical average between two numbers is found by multiplying them and taking the square root of the product. The result is always less than the arithmetic average (i.e., half the sum of the two numbers); and, of course, if one of the numbers is zero, the product is also zero, no matter how large the other number.

be no compansation, any demand however large will, on this basis, be beaten down to zero. There can be no scientific way to fix compensation in *kōgai* cases. What happens in the end is always fixed at the geometrical average.

Finally, I want to talk about the so-called "realistic" way of attacking *kōgai*. Sometimes it is objected that movements based on absolute, uncompromising opposition are nonsense. The Japan Broadcasting Corporation (NHK), run by the government, tells us, for example, that going to court cannot solve the *kōgai* problem; what we need, it asserts, is more "realistic" solutions. But *kōgai* has reached the point where professional engineers like myself, whose business it is to make "realistic" proposals, have given up. I am a man whose job is the treatment of factory waste, but I am unable to reach solutions that way and am now determined to prosecute each and every case of *kōgai* in court.

Only when companies are turned out of many places by unyielding opposition will they begin even reluctantly to undertake some treatment of their waste. This can be seen in some actual examples. In Aichi prefecture there was a pulp factory which was polluting the Kiso river and damaging the paddy fields and culture ponds downstream. When at last it was driven out of the area, it looked for another site. Just when it was about to purchase a new site in Akita prefecture, the local people learned of its past record and refused to let the company come in. It then tried to locate in Onahama, only to have the same thing happen. Eventually it went to Fuji city, already spoiled by pollution. The point is that when people try to think in terms of "realistic" measures, instead of organizing absolute opposition, the result is that many dirty factories descend upon them and stay to pollute.

Now Japanese corporations are beginning to go abroad, to Indochina, Thailand, and so on. We can imagine how they will behave in these countries, how those who have so recklessly ruined cities like Yokkaichi and Minamata in their own country will now go into what will become, in effect, their colonies and destroy them too. Minamata disease will occur, and so will

Yokkaichi asthma. What then happens, with whom will the victims find solidarity? What is our task: to tolerate *kōgai* and work out "realistic" countermeasures, or try to chase the assailant firms out of Japan? I hope that we will be able to say, "We have beaten *kōgai* out of our country this way, and perhaps you will be able to do it the same way."

An Annotated Bibliography of English Language Works on the Social History of Modern Japanese Science

James R. Bartholomew

The social history of modern Japanese science is a relatively new field of scholarship and works available in English are so scarce that the following list may nearly exhaust the literature. As one might expect in a field so new, contributions have come from working scientists and journalists as well as professional historians. Some titles are listed because they contain important information or observations even though their principal focus is not the social history of science in Japan as such. Others are included simply because they are frequently cited or are readily obtainable, even though the quality of their arguments or observations may be low. An effort has been made to state the principal findings of most works and to evaluate their importance whenever this seemed useful.

For those able to read Japanese the following sources are indispensable: *Nihon kagaku gijutsu shi taikei* (Source materials on the history of science and technology in Japan), 25 vols., published by the History of Science Society of Japan from 1964 to 1970. This is the only comprehensive compilation of historical surveys, sources and documents with commentary available on modern science and technology in Japan in any language. See also the Society's journal *Kagakushi kenkyū* (Journal of the history of science, Japan), published quarterly since 1940. The journal presents articles on intellectual and social aspects of science in various parts of the world and covering all time periods. Once a year the History of Science Society of Japan publishes English articles, usually translations from *Kagakushi kenkyū*, in *Japanese studies in the history of science*, Nos. 1–12 (1962–1973).

I. **Modern period (mid-19th century to the present)**

General surveys of institutional aspects of science, including government policy, industrial and private activities, education, etc., include the following;

Bowers, John Z., *Medical education in Japan, from Chinese to Western medicine*. New York, 1965.

A general survey using translated sources and the author's personal observations at the Keio University Medical School.

Fairbank, John, and others, "The influence of modern Western science and technology on Japan and China." *Explorations in entrepreneurial history* (1955), VII: 4, 189–204.

A sophisticated pioneering attempt to delineate the complexities of science's acculturation in Japan (as well as China) by a group of distinguished historians at Harvard. The authors question the utility of "technology" as an analytical concept on grounds of vagueness and lack of specificity yet present their own arguments about science within a framework of equally vague generalizations known as "modernization" theory. Their statement that ideologies, social institutions and political institutions in Japan "proved as much subject to the [corrosive] influence [of science] as [Japan's] military and economic systems" seems doubtful and has been seriously challenged by Craig, among others.

Hashimoto U., "An historical synopsis of education and science in Japan from the Meiji Restoration to the present day." *Impact of science on society* (1963), XIII: 1, 3–25.

Despite the lack of interpretation, this essay by a working scientist is useful mainly because it contains information about institutional aspects of science not readily available elsewhere in English. Particularly valuable are the figures presented on the numbers of scientific societies in Japan between 1868 and 1963 and the extent of their memberships.

Hirosige Tetu, "The role of the government in the development of science." *Cahiers d'histoire mondiale* (1965), IX: 2, 320–339.

A brief but useful description of institutional activities by the Japanese government in science between 1868 and 1965. The author does not, however, discuss the motivations of government policy formulators.

Kamatani Chikayoshi, "The role played by the industrial world in the progress of Japanese science and technology." *Cahiers d'histoire mondiale* (1965) IX: 2, 400–421.

Attempts to explain how industrial research in Japan has been organized since the late 19th century, how it developed in terms both of content and institutional forms, while offering some comparative information on the scale and scope of Japan's research and development effort during the last century with the same endeavors in other industrialized countries. Author states that as late as 1965 even the largest industrial research laboratories in Japan were generally only about one-tenth the size of comparable facilities in leading Western nations. (See also his "The history of research organization in Japan," *Japanese studies in the history of science* (1963) II: 1–79. -Ed.)

Long, T. Dixon, "Science and government in Japan." *Sartryck ur Svensk Naturvetenskap*, Stockholm, 1967, pp. 296–315.

A brief description with some interpretation of the Japanese government's policy toward science during the last century. The author states that the Meiji leaders between 1868 and 1912 "did not succeed in fashioning a full-fledged science policy, but concentrated their efforts on the application of science to major purposes defined by government." He also contends that the absence of a definite program for developing particular disciplines was detrimental to science yet cautions: "Japan's development and absorption of foreign knowledge was so rapid and so ramified throughout the society and economy that it is unlikely . . . government was more than one element . . . in the drama."

Nakayama Shigeru, "A century's progress in Japan's science and technology." *Technical Japan* (1968) I: 1, 77–89.

A brief descriptive analysis of institutional aspects of Japanese research, viewed historically, stressing the importance of private as opposed to governmental initiatives in expenditure of funds for research, a pattern the author notes has distinguished Japan from other advanced industrial nations.

———"Are Japanese only 'good imitators'?" *Technical Japan* (1969) I: 4, 76–81.
Stresses the fact that the institutional structure of modern science created in the late 19th century was supposed to facilitate importation of Western scientific and technical knowledge rather than its creation. Author perhaps overemphasizes the inflexibility of the organizational structure of science as a factor in the way science developed.

———"The role played by universities in scientific and technological development in Japan." *Cahiers d'histoire mondiale* (1965), IX: 2, 340–362.
An informative description and analysis of the relationship of higher education to science in Japan. Author stresses the influence of German models for Japanese universities in the pre-World War II era and on this point should be compared with Bartholomew's analysis (cf. Section II on Early Modern period).

Schwantes, Robert S., "Christianity versus science: a conflict of ideas in Meiji Japan." *Far eastern quarterly* (1953), XII: 2, 123–132.
A brief description of reaction among Protestant missionaries to the introduction of Darwinian evolutionary theories to Japan. Fails to point out rationalistic biases in the indigenous intellectual tradition which ran counter to Christianity but favored science as a factor in the relative failure of the former and the success of the latter.

Tsuge Hideomi, *Historical development of science and technology in Japan*. Tokyo, 1969.

The last three of this book's seven chapters still constitute the only general survey of the history of modern science and technology in Japan available in English. Contains useful facts but little interpretation. Haphazardly written, its value is also diminished by a total absence of footnotes and bibliographical data.

II. **Early Modern period (1868–1920)**

Cultural and institutional information and analyses on scientific research, education, policy, etc., during this period include the following:

Bartholomew, James R., "Japanese culture and the problem of modern science," in Arnold Thackray, ed., *Science and values*. New York, 1974, pp. 92–139.

Analyzes systems of formal and informal organization existing at the Tokyo Imperial University Medical School between 1893 and 1920, making comparisons with the Berlin University Medical School within the context of the German and Japanese university systems of the period. Suggests that any "rigidities" of Japanese medical research and education prior to World War II could not have been due primarily to an inflexible pattern of formal organization as such.

———*The acculturation of science in Japan: Kitasato Shibasaburo and the Japanese bacteriological community, 1885–1920*. Unpublished doctoral dissertation, Stanford University, 1972. (Presently obtainable in hard-cover Xerox form through University Microfilms. -Ed.)

An analysis of relations between the Japanese government and the biomedical science community together with an investigation of patterns of social relationships among biomedical scientists. The study focuses on the Institute of Infectious Diseases and the Tokyo Imperial University Medical School, presenting the argument that factionalism and related "traditional" behavior patterns may well have facilitated rather than impeded the growth of science as is frequently claimed.

Hirakawa Sukehiro, "Changing Japanese attitudes toward Western learning, Part 3." *Contemporary Japan* (1968), XXIX: 1, 138–157.

Contains very interesting and important material on the state of science at Tokyo Imperial University around 1900 as perceived by Dr. Erwin Baelz, a German professor in the University's Medical School, with comments on Baelz's observations by Mori Rintarō, an important figure in Japanese biomedical science and literature of the period.

Lockheimer, Roy F., "Development of science in Japan," in Kalmer H. Silvert, ed., *The social reality of scientific myth*. New York, 1969, pp. 154–170.

Lockheimer's essay succinctly describes the social context of science in Japan which existed in 1868 and the years immediately following. He is further concerned to indicate what lessons the process of establishing science in Japan may have for developing countries of the present day, and therefore states his observations in unnecessarily provocative language, terming, for instance, such phenomena as a common national language a "prerequisite" for science.

Yuasa Mitsutomo, "The scientific revolution and the age of technology." *Cahiers d'histoire mondiale* (1965), IX: 2, 187–207.

Presents the controversial thesis that Japan experienced a scientific revolution in somewhat the same fashion as 17th-century Europe. "Revolution" is used by Yuasa to describe both changes in socialization patterns for scientists (i.e., when were Chinese-style herbalists, mathematicians and physicians replaced by scientists with Western-style training?) and the replacement of the Buddhist and Confucian intellectual systems with belief systems imported from Europe. His thesis is that in all fundamental respects a modern scientific revolution had become institutionalized in Japan by 1889. (See also his "The growth of scientific communities in Japan," *Japanese studies in the history of science*, No. 9 (1970), pp. 137–158. -Ed.)

Materials on the specific sciences in the Early Modern

period are:

Koizumi Kenkichirō, *The development of physics in Meiji Japan, 1868–1012.* Unpublished doctoral dissertation, University of Pennsylvania, 1973. (Presently obtainable in hard-cover Xeros form through University Microfilms. -Ed.)

This pioneering study describes the development of modern physics in Japan during its initial phase, regarded by the author as beginning in 1868 and ending about 1900. Attention is given to both the institutional and the intellectual aspects of science. The author regards 1900 as an approximate turning point because Japan had two centers for producing physicists by that time and because the content of Japanese physical research had attained a genuinely creative level and was attracting attention from the worldwide physical community for the first time.

Yagi Eri, "How Japan introduced Western physics in the early years of the Meiji era, 1868–1888." *Scientific papers of the College of General Education, University of Tokyo* (1959), IX: 1, 163–174.

Yagi points to linguistic differences among the first generation of Japanese physicists, some receiving their physical training in French, some in English, and still others in German, as a major obstacle to the establishment of physics in Japan during the first half of the Meiji period.

For special focus on the interaction between Japan and the outside world, and especially on the impact of the West, see:

Nakamura Takeshi, "The contribution of foreigners." *Cahiers d'histoire mondiale* (1965), IX 2, 294–319.

Presents numerous facts about foreigners who taught in Japan or advised the Japanese on various matters relating to the importation of Western science and technology during the last half of the 19th century. Treatment, however, is sketchy and imprecise with little attempt at interpretation.

Watanabe Masao, "Science across the Pacific: American-Japanese contacts in the late nineteenth century." *Japanese studies in the history of science*" (1970), IX: 115–136.

A highly informative essay on the interactions between American specialists and their Japanese students, with special focus on the introduction of the theory of evolution and its relation to the educational work of Christian missions. The author notes "an almost complete isolation between the scientific-technological aspect and the cultural-religious aspect of life" in this interaction.

Watanabe Minoru, "Japanese students abroad and the acquisition of scientific and technical knowledge." *Cahiers d'histoire mondiale* (1965), IX: 2, 254–293.

The author presents important information not to be obtained elsewhere in English regarding the numbers of Japanese students sent abroad by the government between 1895–1912. Information for this period includes not only where government scholarship student studied but what subjects they pursued. Unfortunately the author presents no information on the latter subject for the period between 1870 and 1895.

Firsthand reports by Western scientists involved in the early stage of introducing modern science in Japan include:

Baelz, Erwin, *Awakening Japan: the diary of a German doctor.* New York, 1932.

Contains important observations on the state of science in Japan around the turn of the century by a one-time professor of the Tokyo Imperial University Medical School. Readers should compare Baelz's remarks with the analysis of his views and motives presented by Hirakawa Sukehiro above.

Morse, Edward S., *Japan day by day*, 2 vols. Boston and New York, 1918.

Morse was most probably the first and certainly the most influential scientist to introduce Darwin's evolutionary theory to Japan as professor of zoology at Tokyo University (1877–79). Unfortunately his comments on the state of science and his descriptions of the University are few but nevertheless worth examining. He notes, for instance, how great was the Japanese expenditure on education about 1880 as compared with the effort then being made in Russia. (For comparison with a lesser known and less influential evolutionist, see Watanabe Masao, "John Thomas Gulick: American evolutionist and missionary in Japan," *Japanese studies in the history of science*, No. 5 (1966), pp. 140–149, as well as this author's essay listed above. -Ed.)

III. **The Prewar period (1920's to early 1940's)**

Virtually nothing on this period has been available in English prior to the publication of this book, *Science and society in modern Japan*. Taketani Mituo's paper (I-2) for this volume was adapted from an English article in the *Supplement* of the *Progress of theoretical physics*, No. 50, 1971, and Hirosige Tetu's study (III-3) appeared in an early issue of *Japanese studies in the history of science*, No. 2 (1963). Other papers in this volume which help fill this gap are: I-3; II-1; and III: 1, 2, 3, 4, 5,

IV. **Postwar period (late 1940's to the present)**

Though fewer than one might expect for this crucial period, there are, in addition to papers I-4, III-6 and III-7 in this volume, the following:

Boffey, Philip M., "Japan (I): on the threshold of an age of big science?" *Science* (1970), 167: 3914, 31–45,
———"Japan (II): university turmoil is reflected in research." *Science* (1970), 167: 3915, 147–152.
———"Japan (III): industrial research struggles to close the gap." *Science* (1970), 167: 3916, 264–267.

A remarkably insightful analysis of the postwar research system, by a non-specialist, which neatly balances the elucidation of apparent strengths and weaknesses in Japanese science.

Campbell, Louise, "Science in Japan." *Science* (1964), 143: 3608, 776–782.

A brief survey of the institutional growth of Japanese science between 1945 and 1964.

Long, T. Dixon, "Policy and politics in Japanese science: the persistence of a tradition." *Minerva* (1969), VII: 3, 426–453.

The best discussion of the relationship of science to government of the postwar era available in English. The author contends that among the reasons for the relatively backward state of science policy as such in Japan are the political preferences of the academic and scientific communities for the left-wing parties and their hostility to the ruling Liberal Democratic Party, the dominance of the civil service in the legislative field the disinclination of many academic research scientists to concern themselves with the practical applications of their work. Illuminating comparisons of research and development expenditures and the functioning of science policy in Japan with practices in the other industrial nations adds to the value of Long's analysis. (Mr. T. Dixon Long served as Rapporteur for the Organization for Economic Co-operation and Development (OECD) Examiner's Report, *Reviews of National Science Policy: Japan*, Paris, 1967, a detailed but rather bureaucratic compilation which assumes that greater centralization would serve national science policy better. -Ed.)

MacKay, Alan, "An outsider's view of science in Japan," *Impact of science on society* (1962), XII: 3, 177–201.

An impressionistic series of observations by a British physicist, not limited strictly to scientific matters. Contains some useful information on levels of funding for research and development in several advanced industrial nations, pointing up the low percentage of national income allocated for these purposes by Japan as of 1961 compared to that of Britain, the United States or the Soviet Union.

Supreme Command of the Allied Powers (SCAP), General Headequarters, *Japanese natural resources*. Tokyo, 1949. Cf. Capter 12, "Scientific research and technical competence in relation to resource utilization," pp. 503–519.

A general survey of conditions existing in the early postwar years, with interesting interpretive comments interspersed. Authors argue, for example, that "Japanese scientists in the past have exhibited shortcomings in cooperation, coordination and criticism to a degree that distinguishes them from British or American scientists." Some examples are provided as to behavior patterns among scientists. Report is naturally biased in favor of an organizational or institutional pattern for science like that then existing in the United States.

Yoshida Tomizo, "Organization of scientific activities," in Arthur Livermore, ed., *Science in Japan*. Washington, D.C., 1965, pp. 1–24.

Consists mostly of tables indicating all laboratories and institutes supported by the national government with the numbers of researchers at each as of 1963. Some gross information regarding research and development expenditure by private firms is also presented.

V. **Biographical data**

The only general source in English is:

Iseki Kuro, ed., *Who's who: hakushi in great Japan*, 5 vols. Tokyo, 1921–1930.

An indispensable source for anyone interested in the social history of modern Japanese science, this remarkable work, published in English and Japanese together, contains biographical sketches of all Japanese receiving the doctorate (*hakushi-go*) in all fields of knowledge prior to 1930. For the natural sciences, 240 doctorates were awarded between 1888 and 1930 in physics and chemistry, 2,072 in medical science, and 453 in engineering. Iseki saw fit to include not only information on the professional achievements of this top echelon of

Japanese scientists but data on their social class c
tion in the birth order of siblings and education b
as well. (Some data can also be gleaned from Yuka
ed., *Profiles of Japanese science and scientists*, Tokyo, 1970. -Ed.),

The only complete monograph on an individual available in English, so far, is;

Eckstein, Gustav, *Noguchi*. New York, 1931.
A biography of a prominent Japanese bacteriologist who became deputy director of the Rockefeller Institute. Although Noguchi did most of his work in the United States, he began in Japan and returned there in 1915, circumstances which permitted the author to report on some significant features of the social context of Japanese biomedical research characteristic of the time.

VI. **Premodern period (prior to 1868)**

Though beyond the scope of the present volume, some readers may wish to consult materials on the premodern background of Japanese science, particularly those treating early Western influences on the transition to modern science in Japan:

Bowers, John Z., *Western medical pioneers to feudal Japan*. Baltimore and London, 1970.
This work focuses primarily on the work of Dutch and other European physicians in Japan in the late 18th and early 19th centuries; the last chapter also presents an exellent discussion of the Japanese decision in 1870 to adopt German medicine as the model for its own system of modern medical training, together with the factors which motivated that decision. (N.B. The author's competence in the Dutch language affords English readers some valuable information on the Western medical personnel in Japan in the period, but data on Japan sometimes needs confirmation or correction from Japanese sources. -Ed.)

Craig, Albert M., "Science and Confucianism in Tokugawa Japan," in Marius Jansen, ed., *Changing Japanese attitudes toward modernization*. Princeton, 1965, pp. 133–160.

A brilliant, incisive pioneering analysis of a strategic part of the intellectual setting pertinent to science which existed prior to the Restoration of 1868. Contends that through the concept of *ri* the orthodox Neo-Confucian tradition rationalized the acceptance of Western science as a legitimate form of knowledge from a Confucian point of view, and suggests as well that science's introduction to Japan had a qualitatively different impact on culture and society than characterized its rise to intellectual prominence in Europe. Readers should note Craig's retraction in a later essay (in Robert Ward, ed., *Political development in modern Japan*) of the claim made here that Nishi Amane accomplished the "final destruction of the Chu Hsi concept of *ri*," observing that for many less sophisticated early Meiji thinkers an undifferentiated notion of *ri* covering science and ethics remained acceptable for some time thereafter.

Goodman, Grant K., *The Dutch impact on Japan, 1640–1853.* Leiden, 1967.

Though emphasizing intellectual aspects of science in Japan prior to the Restoration, this interesting book also analyzes the social and cultural context within which the importation of science took place. Goodman contends, for instance, that urbanization during the Tokugawa period (1600–1867) was indispensable for the flourishing of science which occurred during the same period. His conclusions about the impact science had on society and culture before 1868 differ sharphy from those of Craig, it should be noted.

Nakayama Shigeru, *A history of Japanese astronomy: Chinese background and Western impact.* Cambridge, Massachusetts, 1969.

Though primarily an intellectual history of premodern development in Japanese astronomy, this work neatly summarizes the social and cultural reasons for the limited creativity of the exact sciences in Japan before the Restoration and presents important information on the institutionalization of Western astronomy in the early post-Restoration years. The author's discussion of the discontinuities between pre-Restoration and post-Restoration science is particularly valuable for students of the social history of modern science in Japan.

Otori Ranzaburō, "The acceptance of Western medicine in Japan." *Monumenta Nipponica* (1964), XIX: 3, 20–40.

General survey of medical history from the mid-16th century to the beginning of the Meiji era (1868), emphasizing European medical texts with which Japanese became familiar. Stresses continuity between pre-Restoration and post-Restoration developments in the intellectual aspects of medical science. (N.B. This special issue of *Monumenta Nipponica* gives general surveys of the adoption of most Western sciences and some technologies, though not always with data or interpretation of the social history involved.- Ed.)

Index

J

K

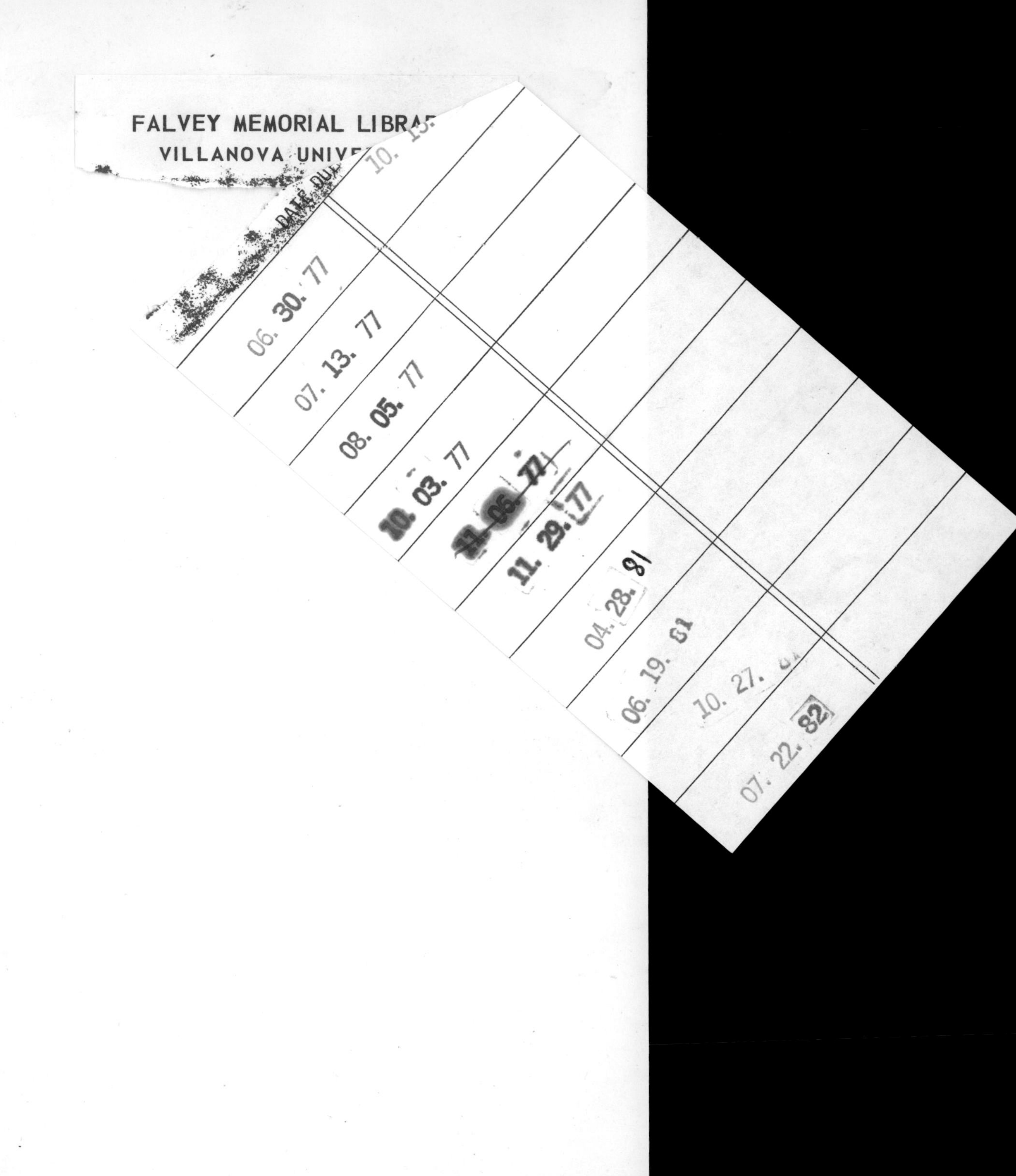